Perspectives in Neural Computing

Springer
*London
Berlin
Heidelberg
New York
Barcelona
Budapest
Hong Kong
Milan
Paris
Santa Clara
Singapore
Tokyo*

Also in this series:

J.G. Taylor
The Promise of Neural Networks
3-540-19773-7

Maria Marinaro and Roberto Tagliaferri (Eds)
Neural Nets - WIRN VIETRI-96
3-540-76099-7

Adrian Shepherd
Second-Order Methods for Neural Networks: Fast and Reliable Training Methods for Multi-Layer Perceptrons
3-540-76100-4

Jason Kingdon
Intelligent Systems and Financial Forecasting
3-540-76098-9

Dimitris C. Dracopoulos

Evolutionary Learning Algorithms for Neural Adaptive Control

Springer

Dimitris C. Dracopoulos, BSc, MSc, PhD, DIC
Department of Computer Science, Brunel University,
Uxbridge, Middlesex UB8 3PH, UK

Series Editor

J.G. Taylor, BA, BSc, MA, PhD, FInstP
Centre for Neural Networks,
Department of Mathematics, Kings College,
Strand, London WC2R 2LS, UK

ISBN 3-540-76161-6 Springer-Verlag Berlin Heidelberg New York

British Library Cataloguing in Publication Data
Dracopoulos, Dimitris C.
  Evolutionary learning algorithms for neural adaptive
  control. - (Perspectives in neural computing)
  1.Neural networks (Computer science)   2.Genetic algorithms
  3.Adaptive control systems
  I.Title
  006.3'2
ISBN 3540761616

Library of Congress Cataloging-in-Publication Data
A catalog record for this book is available from the Library of Congress

Apart from any fair dealing for the purposes of research or private study, or criticism or review, as permitted under the Copyright, Designs and Patents Act 1988, this publication may only be reproduced, stored or transmitted, in any form or by any means, with the prior permission in writing of the publishers, or in the case of reprographic reproduction in accordance with the terms of licences issued by the Copyright Licensing Agency. Enquiries concerning reproduction outside those terms should be sent to the publishers.

© Springer-Verlag London Limited 1997
Printed in Great Britain

The use of registered names, trademarks etc. in this publication does not imply, even in the absence of a specific statement, that such names are exempt from the relevant laws and regulations and therefore free for general use.

The publisher makes no representation, express or implied, with regard to the accuracy of the information contained in this book and cannot accept any legal responsibility or liability for any errors or omissions that may be made.

Whilst we have made considerable efforts to contact all holders of copyright material contained in this book, we have failed to locate some of these. Should holders wish to contact the Publisher, we will be happy to come to some arrangement with them.

Typesetting: Camera ready by author
Printed and bound at the Athenæum Press Ltd., Gateshead, Tyne and Wear
34/3830-543210 Printed on acid-free paper

To my parents and my brother

Preface

This book is the result of a part of my research and teaching during the last few years. It concerns the use of neural networks and genetic algorithms for advanced control applications, in relation to dynamic systems theory. It was written with a wide audience in mind:

- Researchers working with control and adaptive control problems.

- Professionals, researchers, students wishing to learn the fundamentals of neural networks and genetic algorithms in the framework of real-life applications.

- Experienced control theorists not familiar with neural network and evolutionary techniques for solving high–nonlinear control theory problems.

- Neural network and genetic algorithms researchers who would like to extend their field of applications to dynamic systems, chaos and control.

- Researchers working with dynamic systems.

- Postgraduate students in engineering and computer science.

I have tried to cover as much of the above as possible, and the reader will find in the bibliography many relevant publications for further reading. The two chapters on neural networks and genetic algorithms can be used as a fast route to obtaining the necessary knowledge for their application, not necessarily for control applications or dynamic systems. However, the rest of the book was written from a "control" point of view. I hope that the book will serve its aims and purpose for all categories of readers.

I would like to thank Professor Antonia J. Jones, of the University of Wales, Cardiff, for her great help and comments on most of this work, and the late Professor Patrick Parks of the Oxford Mathematical Institute, for his valuable comments on parts of the book. The book was finished while I was in a one-year professor position at Ecole Normale Supérieure de Lyon, Laboratoire de l'Informatique du Parallélisme, France, being on a sabbatical leave from Brunel University, in London. Therefore, thanks go to both institutes. Thanks to

Professor John Taylor, editor of the series, for his suggestions for improvements in the book contents, and thanks to Rosie Kemp, assistant editor of the Springer series, for her helpful editing tips.

Finally but above all, I would like to thank my parents and my brother Mike. Without them, it would have been impossible not only to finish this book, but even to start it.

Dimitris C. Dracopoulos
London, May 1997

Contents

1	**Introduction**	**1**
2	**Dynamic Systems and Control**	**5**
	2.1 Systems Theory	5
	2.2 Nonlinear Dynamic Systems and Chaos	6
	2.3 Control	11
	2.4 Traditional Adaptive Control Methods	13
	2.4.1 The Self-Tuning Regulator (STR)	14
	2.4.2 The Model-Reference Adaptive Control (MRAC)	16
	2.4.3 Limitations	18
	2.5 The OGY Method for the Control of Chaotic Systems	19
3	**The Attitude Control Problem**	**23**
	3.1 The Euler Equations	26
	3.2 Chaos in the Euler Equations	30
	3.3 Orientation of a Rigid Body	31
	3.4 Defining the Problem	37
	3.5 Attitude Control Approaches	38
	3.6 Numerical Methods for Solving Ordinary Differential Equations	42
	3.6.1 The Runge-Kutta Method	43
	3.6.2 The Adams-Moulton Method	45
	3.7 The Attitude Control Simulator	46
4	**Artificial Neural Networks**	**47**
	4.1 Multilayer Feedforward Neural Networks	48
	4.2 The Backpropagation Algorithm	50

    4.3 Other Neural Network Paradigms 52
           4.3.1 Hopfield Networks 52
           4.3.2 Boltzmann Machines and Simulated Annealing 58
           4.3.3 Unsupervised Competitive Neural Networks 61
           4.3.4 Adaptive Resonance Theory (ART) Networks 64
           4.3.5 Radial Basis Function Networks 68

5 Neuromodels of Dynamic Systems   71

    5.1 A Pattern Recognition Task 74
    5.2 Local Predictive Networks (LPN) Applied to Dynamic
          System Modeling 78
           5.2.1 LPN for a Two Point Attractor System 79
           5.2.2 LPN for the Van der Pol Equation 82
           5.2.3 LPN Applied to a Chaotic System 86
           5.2.4 LPN for the Euler Equations 89
    5.3 The Importance of Modeling to Adaptive Control 93

6 Current Neurocontrol Techniques   97

    6.1 Supervised Control 98
    6.2 Direct Inverse Control 99
    6.3 Neural MRAC Adaptive Control 102
    6.4 Reinforcement Learning Control 103
    6.5 A General Framework for Adaptive Control 106

7 Genetic Algorithms   111

    7.1 Genetic Algorithms 112
    7.2 The Genetic Algorithm Underlying Idea 116
    7.3 Classifier Systems 118
    7.4 Genetic Programming 119
    7.5 Genetic Programming for the Detumbling of a Rigid Body
          Satellite 123
    7.6 The Design of a Genetic Algorithm for the Attitude
          Control Problem 127

8 Adaptive Control Architecture   133

    8.1 Introduction 133
    8.2 The Adaptive Control Architecture 134
    8.3 The Use of LPN for the Adaptive Control Architecture 135

8.4	Spin Stabilization of a Satellite About a Stable Axis	139
8.5	Spin Stabilization of a Satellite About an Unstable Axis	143
8.6	Spin Stabilization of a Satellite Subject to Sensor Noise	147
8.7	Satellite Attitude Control Subject to External Forces Leading to Chaos	151
8.8	Control Subject to Sensor Noise and External Forces Leading to Chaos	155
8.9	An Extension to the Adaptive Control Architecture	159
8.10	Discussion	161

9 Conclusions and the Future — 165

A Euler Equations Solutions — 169
- A.1 An Analytical Solution for Euler Equations ... 169
  - A.1.1 Special Case I ... 169
  - A.1.2 Special Case II ... 173
- A.2 Computing a Solution for Euler Equations ... 174

B An Attitude Control Simulator — 179

Bibliography — 187

Index — 207

Chapter 1

Introduction

"I'll tell it", said the Mock Turtle, in a deep hollow tone, "sit down, and don't speak till I've finished".
So they sat down, and no one spoke for some minutes: Alice thought to herself "I don't see how it can ever finish, if it doesn't begin", but she waited patiently.

Lewis Carroll
(Alice's Adventures Under Ground)

Control in general, and automatic control in particular, are of major importance in modern society. Systems from home appliances and automobiles, to robots, spacecrafts and nuclear reactors, require automatic control systems to maintain their operation. A control system may be required to maintain a stable condition, to follow a pre-specified path or to produce a periodic action.

Today, most controllers are simple in design and fixed at the time of operation. They suffer from limitations in the range of operation, improper operation, inefficiency due to their simple design (which often makes many assumptions), and lack of adaptability. The rising complexity of dynamic systems, coupled with increasingly stringent performance criteria, necessitates the utilization of more complex and sophisticated controllers. These controllers must be robust to faults and disturbances, able to cope with nonlinear relationships and model uncertainties, and make provision for temporal and spatial variations of the system [149].

Automatic control researchers have developed specialized algorithms to incorporate more knowledge into controllers. However, much of the work is narrow in focus and has been concentrated on specific solutions to precisely defined problems [12].

Many control tasks continue to be solved using very simple controllers. An example is the process control industry. Most of its systems are still implementing PID (proportional-integral-derivative) controllers, a technique which has been in use for many decades. A problem with new techniques is that they require specialized knowledge in order to be properly implemented. What is needed are more general techniques, which are more cost effective to use, and which provide improved control over the simple controllers. This can be done by building controllers that initially contain sufficient knowledge, and are adaptable to the changes of the environment and the variations in the actual system dynamics [148].

Some recent research on neural networks emulating simple classical control problems has shown encouraging results [7, 17]. The main purpose of this book is to investigate how neural networks and genetic algorithms can be applied to difficult adaptive control problems which conventional control methods are either unable to solve, or do not typically give acceptable results. The book presents both a technique of modeling complex dynamic systems and a general control architecture using neural networks and genetic algorithms, which addresses the problem of adaptive control by using minimum initial knowledge of controller design, and learning from the process through time. The method does not require knowledge of the plant dynamics in analytic form, but is adaptive to unknown changes in the plant, by observing its input-output behavior. The resulting controller is general in nature, and provides self-adaptive nonlinear control without linearization of any system component.

One of the principal motivations for this work is the observation that the neural systems of living organisms perform such functions in an extraordinarily capable manner, and appear to avoid the limitations of traditional control methods. Neural computation is also an attractive approach because, at least in principle, very high speed implementations can be constructed. This opens up the possibility of applications in which the control system is capable of real-time adaptation to changes in the physical dynamics of the system.

Thus, the objectives of this work are:

First, to develop a technique for modeling complex dynamic systems. Given this capability, a plant model can be constructed every time that the system

CHAPTER 1. INTRODUCTION

characteristics are changed or the system dynamics are unknown.

The second objective is to develop an adaptive controller which maintains or drives the plant in a desired operation. The design objectives for the adaptive controller are: control of difficult highly non-linear dynamic systems (including chaotic systems), operation in real-time applications (at least in principle), prediction capability, robustness, stability, and goal-directed control.

The book considers the control of highly nonlinear physical systems with unknown parameters. The adaptive control architecture described is evaluated by computer simulation. As a testbed the adaptive problem of the automatic detumbling of a satellite is used. The book investigates the feasibility of applying neural network and genetic algorithm architectures to the control of nonlinear systems and therefore it is aimed at exposing the *principles* involved, rather than modeling specific real systems, such as a satellite. In choosing the dynamic system to be modeled a number of criteria were considered. The system to be modeled was chosen to represent a large class of realistic control applications and to exhibit complex highly non-linear behavior (and sometimes chaotic behavior) such as to pose problems for current conventional control system methodologies. The book is not necessarily intended for people familiar with neural networks and genetic algorithms techniques. Thus, two chapters are devoted to an introduction to the fundamental principles of neural networks and genetic algorithms.

The goal of the specific control application considered here as a testbed is to detumble the satellite (reduce its kinetic energy) while achieving a desired orientation (attitude control problem). It is assumed that the satellite is a rigid body, so that its motion is described by the Euler equations [74]. It is supposed that the system is equipped with reaction thrusters which provide control torques about the three principal axes of inertia. For most of the time it will be assumed that no external forces (besides the control torques) are acting upon the system.

The above control problem has no complete general solution [139], and is considered as a difficult problem in the control bibliography. It is highly nonlinear, exhibits chaotic behavior under certain circumstances [126], and is therefore a challenging problem for testing the described neuro-genetic control architecture.

Chapter 2 is an overview of dynamic systems and control. It includes system theory, properties of linear, nonlinear and chaotic systems, and conventional adaptive control techniques. A description of the OGY method, a "conventional" method developed for the control of chaotic dynamic systems is also given. The attitude control problem described by the Euler equations is developed in Chapter 3. For completeness the classical equations for the rotational motion of a rigid body and its orientation are derived. It is also shown that under certain circumstances the system can exhibit chaotic behavior. Following this, the attitude control problem as viewed in this book is defined, and a literature review of the problem is given. The remainder of Chapter 3 concentrates on numerical methods for solving systems of ordinary differential equations.

Chapter 4 focuses on Artificial Neural Network techniques and their properties, with emphasis on feedforward networks and the back-propagation algorithm. A description of Hopfield, Boltzmann machines, Kohonen, adaptive resonance theory (ART) and radial basis function networks is also included. In Chapter 5, an algorithm for modeling different dynamic systems using neural networks is described, and it is applied to dynamic systems of increasing complexity (including two chaotic systems). Chapter 6 studies current techniques for developing neural network controllers, and outlines a general framework for adaptive control using neural networks.

Chapter 7 introduces a family of evolutionary algorithms, specifically genetic algorithms, genetic programming and some of their properties. The theoretical stability analysis of a control solution, found by genetic programming for the detumbling of a rigid body satellite is detailed. A genetic algorithm for the attitude control problem is described and simulation results for this algorithm are included. The use of genetic algorithms in learning as classifier systems is also included. Chapter 8 describes a recent hybrid (neural networks and genetic algorithms) adaptive control architecture and presents some simulation results for the Attitude Control problem. The chapter closes by examining a possible extension to the neuro-genetic adaptive control architecture presented. Chapter 9 presents suggestions for how neurocontrollers and hybrid control methods can be improved and discusses the future of such methods. Appendix A describes the analytical solution of Euler equations for various special cases. Finally, Appendix B includes the ANSI C code for an attitude control simulator.

Chapter 2

Dynamic Systems and Control

Fortis est veritas

Most of the observed phenomena in every day life have dynamic components. The term *dynamic* refers to phenomena that produce time-changing patterns, the characteristics of the pattern at one time being interrelated with those at other times [131]. The term is nearly synonymous with time-evolution or pattern of change. It refers to the unfolding of events in a continuing evolutionary process. Examples from the daily life include home heating systems, the development of the economy, and population growth.

2.1 Systems Theory

A *system* may be broadly defined as an aggregation of objects united by some form of interaction or interdependence. When one or more aspects of the system change with time, it is generally referred to as a *dynamical system*.

The principal concern of systems theory is the behavior of systems deduced from the properties of subsystems or elements of which they are composed, and their interaction. Influences that originate outside the system and act upon it, without being directly affected by what happens in the system, are called *inputs*. The quantities of interest that are affected by the action of these external influences, are called *outputs* of the system. As a mathematical concept, a dynamical system can be considered as a structure that receives an input $u(t)$ at each time instant t, where t belongs to a time set $\mathcal{T}$, and produces an output $y(t)$. The values of the input are assumed to belong to some set $\mathcal{U}$, while those of the output belong to a set $\mathcal{Y}$. In most cases the output $y(t)$ depends not only on $u(t)$, but also on the past history of the inputs, and hence that of the system. The state of the dynamic system at time t is denoted by $x(t)$, and its elements $x_i(t) (i = 1, 2, \ldots, n)$ are called *state variables*. The state $x(t)$ at time t is determined by the state $x(t_0)$, for any time $t_0 < t$, and the input u defined over the interval $[t_0, t)$.

An *adaptive system* is a system which is provided with a means of continuously monitoring its own performance in relation to a given figure of merit or optimal condition and a means of modifying its own parameters by a closed loop action so as to approach this optimum [148]. Adaptive systems are inherently nonlinear [131]. Their behavior is therefore quite complex, which makes them difficult to analyze. Progress in theory has been slow, and much work remains before a reasonably complete, coherent theory is established. Because of the complex behavior of adaptive systems, it is necessary to consider them from several points of view. Theories of nonlinear systems, stability, system identification, recursive parameter estimation, optimal control, stochastic control and recently neurocontrol, contribute towards the understanding of adaptive systems.

2.2 Nonlinear Dynamic Systems and Chaos

Dynamic systems are represented mathematically, in terms of either differential or difference equations. These equations provide the structure for representing time linkages among variables.

The use of either differential or difference equations to represent dynamic behavior corresponds respectively to whether the behavior is viewed as occurring in continuous or discrete time. Dynamic behavior described by differential equations relates the derivatives of a dynamic variable to its current value. Discrete time consists of an ordered sequence of points rather than a continuum. *Difference equations* relate the value of a variable at one time to the values at adjacent times.

For control applications, the distinction between linear and nonlinear dy-

2.2. NONLINEAR DYNAMIC SYSTEMS AND CHAOS

namic systems is important. A linear system can be defined as one which is described by a linear differential equation[1]. Similarly a system described by a nonlinear differential equation, is called a *nonlinear dynamic system*.

The vast majority of physical systems are nonlinear dynamic systems. Most work in the domain of automatic control has concentrated on linear systems, mainly because of a lack of efficient tools in the analysis of nonlinear systems. At first sight, this is not a drawback, since we can linearize a nonlinear system about a point. However, this assumption is valid only as long as the system's operation remains sufficiently close to the state about which equations were linearized.

Even first-order nonlinear dynamic systems can exhibit forms of behavior that are drastically different from those in linear systems. Thus linearized models of dynamic systems are not suitable for many applications. Mathematically, the essential difference between linear and nonlinear systems is clear. Any two solutions of a linear equation can be added together to form a new solution, according to the *superposition principle*. In fact, superposition is responsible for the systematic methods used to solve (independent of other complexities) almost any linear problem. Fourier and Laplace transform methods, for example, depend on being able to superpose solutions. One can break the problem into smaller subproblems and then add the separate solutions to get the solution of the whole problem.

In contrast, two solutions of a nonlinear system cannot be added together to form another solution. Superposition fails. Thus, a nonlinear system has to be considered as a whole; one cannot (at least not obviously) break the system into smaller subproblems and then add the solutions. It is therefore not surprising that no general analytic approach exists for solving typical nonlinear equations.

The dynamics of simple (in mathematical form) nonlinear systems can be highly complex and this can be the result of the presence of nonlinearity in the equations, instead of the complex interaction of many differential equations. For instance, when water flows through a pipe at low velocity, its motion is *laminar*, and is characteristic of linear behavior. It is regular, predictable, and describable, in simple analytic mathematical terms. However, when the velocity exceeds a critical value, the motion becomes *turbulent*, with localized eddies moving in a complicated, irregular, and erratic way that typifies nonlinear behavior [82], [145]. Using this and other examples, we can isolate at least

---

[1]Since most of the dynamic systems that this book deals with are in continuous time, definitions will be given in terms of this domain.

three characteristics that distinguish linear and nonlinear physical phenomena [31]:

1. The motion itself is qualitatively different. Linear systems typically show smooth, regular motion in space and time that can be described in terms of well-behaved functions. Nonlinear systems, however, often show transitions from smooth motion to chaotic, erratic, or even apparently random behavior.

2. The response of a linear system to small changes in its parameters or to external stimulation is usually smooth and in direct proportion to the stimulation. But for nonlinear systems, a small change in the parameters can produce an enormous qualitative difference in the motion. Furthermore, the response to an external stimulation can be different from the stimulation itself. For example, a periodically driven nonlinear system may exhibit oscillations, at say, one-half, one quarter, or twice the period of the stimulation.

3. A localized "lump", or pulse, in a linear system will normally decay by spreading out as time progresses. This phenomenon, known as *dispersion*, causes waves in linear systems to lose their identity and die out, such as when low-amplitude water waves disappear as they move away from the original disturbance. In contrast, nonlinear systems can have highly coherent, stable localized structures (such as the eddies in turbulent flow) that persist either for long periods of time, or in some idealized mathematical models, for all time. The remarkable order reflected by these persistent coherent structures, stands in sharp contrast to the irregular, erratic motion that they themselves can undergo.

One of the remarkable phenomena encountered in nonlinear systems is that of *chaos*. Deterministic chaos is the term applied to the aperiodic, irregular, unpredictable, apparently random behavior in the time evolution of certain nonlinear dynamic systems. In the last two decades, chaos has been observed in an incredible variety of nonlinear mathematical models and natural phenomena [14], [72], [193]. Although the processes are strictly deterministic and all forces are known, extreme sensitivity to initial conditions is observed, creating the impression of a random behavior to an external observer.

In the real world exact knowledge of the initial state is not achievable. No matter how accurately the velocity of a particular particle is measured,

2.2. NONLINEAR DYNAMIC SYSTEMS AND CHAOS

one can request more accuracy. Despite this, it is usually assumed that if the initial conditions of two separate experiments are almost the same, the final states will be almost the same. For most smoothly behaved, "normal" systems, this assumption is correct. But for many nonlinear systems it is false, and deterministic chaos is the result. For these reasons, in current literature chaos is a term synonymous with the *sensitive dependence on initial conditions* [177], [178].

An example of chaotic behavior in nonlinear systems is that of Lorenz equations. This model was developed in the early 1960's by E. Lorenz, a meteorologist who was convinced that the unpredictability of weather forecasting was not due to any external noise or randomness but was in fact compatible with a completely deterministic description. In this sense, Lorenz was attempting to make precise the qualitative insight of Poincaré who, in his *Science and Method*, mentioned a weather sensitive dependence on initial conditions. To demonstrate this sensitive dependence, Lorenz proposed a model for thermally induced fluid convection in the atmosphere. Fluid heated from below becomes lighter and rises, whereas heavier fluid falls under gravity. Such motions often produce convection rolls similar to the motion of fluid in a circular torus. His model is described by the following system of ordinary differential equations:

$$\begin{aligned} \dot{x} &= \sigma(y-x) \\ \dot{y} &= \rho x - y - xz \\ \dot{z} &= xy - \beta z \end{aligned} \qquad (2.1)$$

The variable x is proportional to the amplitude of the fluid velocity circulation in the fluid ring, while y and z measure the distribution of temperature around the ring. The parameters σ and ρ are related to the Prandtl number (describing the relative strength of viscous to thermal forces) and the Rayleigh number (giving the strength of the thermal driving) respectively. The β parameter is a geometric factor.

For $\sigma = 10$ and $\beta = 8/3$ (a favorite set of parameters for experts in the field), there are three equilibria for $\rho > 1$ for which the origin is an unstable saddle point [192]. When $\rho > 25$, the other two equilibria become unstable spirals and a complex chaotic trajectory moves between all three equilibria (Figure 2.1).

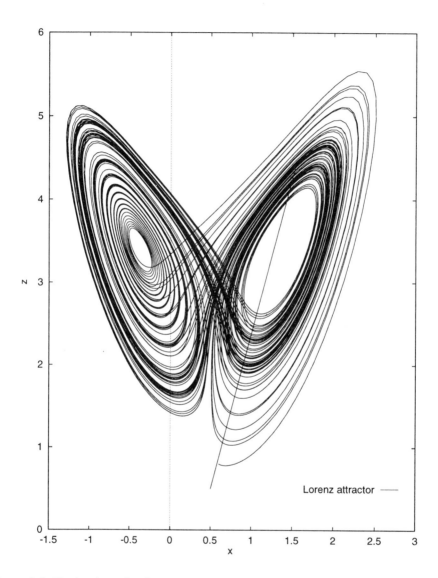

Figure 2.1: Projection of a chaotic trajectory of a solution to the Lorenz equations in the $x - z$ plane.

2.3 Control

Although the study of a particular dynamic situation is sometimes motivated by the simple philosophic desire to understand the world and its phenomena, many analyses have the explicit motivation of devising effective means for changing a system, so that its behavior pattern is in some way improved. The means for affecting behavior can be described with the term *control*.

Control of a process means qualitatively, the ability to direct, alter, or improve its behavior, and a control system is one in which some quantities of interest are maintained accurately around a prescribed value. Control becomes truly automatic when systems are made to be self-regulating. This is done by introducing the concept of *feedback* which is one of the fundamental ideas of engineering. The essence of the concept consists of the triad: *measurement*, *comparison* and *correction*. By measuring the quantity of interest, comparing

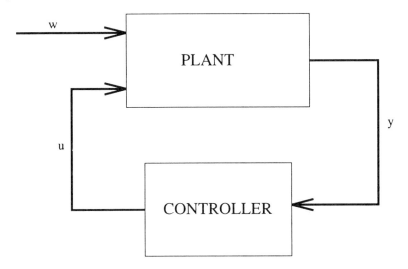

Figure 2.2: Feedback control.

it with the desired value, and using the error to correct the process, the familiar chain of cause and effect in the process is converted into a closed loop of interdependent events. The process to be controlled is called the *plant* and interacts with its environment by means of two types of input and output channels (Figure 2.2). The signal w contains all inputs that cannot be directly affected by the controller. This includes disturbances as well as signals to be tracked and other reference information. The signal y, which is the measured output of the plant, contains all data which are available to the controller, such as the values of the state variables. The control signal u is the part of the plant input which can be manipulated. This closed sequence of information transmission, referred to as feedback, underlies the entire technology of automatic control based on self-regulation.

One of the primary purposes of classical feedback control is to increase the robustness of a control system, i.e. to increase the degree to which the system performs well when there is uncertainty. Classical linear control provides robustness over a relatively small range of uncertainty.

There are many design techniques for generating control strategies when the model of the system is known. When the model is unknown, on-line parameter estimation could be combined with on-line control. This leads to adaptive, or self-learning, controllers . The basic structure of an adaptive controller is shown in Figure 2.3.

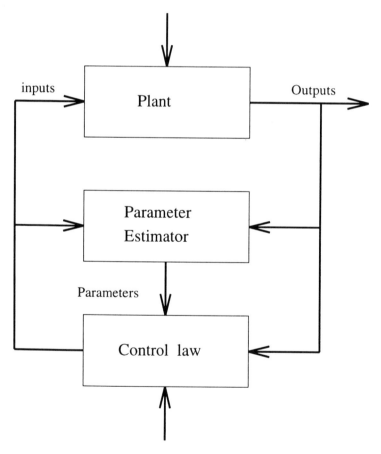

Figure 2.3: Basic structure of an adaptive controller.

An *adaptive controller* can therefore be defined as a feedback regulator that can modify its behavior in response to changes in the dynamics of the process and the disturbances, so as to operate in an optimum manner according to some specified criterion. Adaptive control techniques have been developed for systems that must perform well over large ranges of uncertainties due to large variations in parameter values, environmental conditions and signal inputs. These adaptive techniques generally incorporate a second feedback loop, which is outside the first feedback loop. This second loop may have the capability to track system parameters, environmental conditions, and input characteristics. Then feedback control may vary parameters in compensation elements of the inner loop to maintain acceptable performance characteristics.

The objective of the design of an *intelligent control system* is similar to that of the adaptive control system. However, there is a difference. For an intelligent control system, the range of uncertainty may be substantially greater than can be tolerated by algorithms for adaptive systems. The main objective with intelligent control is to design a system with acceptable performance characteristics over a very wide range of uncertainty [18].

Generally, complex systems are characterized by poor models, high dimensionality of the decision space, distributed sensors and decision makers, high noise levels, multiple performance criteria, complex information patterns, overwhelming amounts of data and stringent performance requirements. We can broadly classify the difficulties that arise in these systems into three categories for which established methods are often insufficient [149]. The first is computational complexity, the second is the presence of nonlinear systems with many degrees of freedom, and the third is uncertainty. The third category includes model uncertainties, parameter uncertainties, disturbances and noise. The greater the ability to deal successfully with the above difficulties, the more intelligent the control system is. Qualitatively, a system which includes the ability to sense its environment, process the information to reduce the uncertainty, plan, generate and execute control actions, constitutes an intelligent control system.

2.4 Traditional Adaptive Control Methods

In this section two of the most well known traditional adaptive control techniques are briefly described. Today most control theorists do not make a distinction between these two approaches, since they are in fact very similar. The description here follows the historical development and thus it treats the Self-Tuning Regulator method in discrete time. More details and extensions of these methods can be found in [13], [76], [148].

2.4.1 The Self-Tuning Regulator (STR)

In an adaptive system it is assumed that the regulator parameters are adjusted all the time. This implies that the regulator parameters follow changes in the process. However, it is difficult to analyze the convergence and stability properties of such systems. To simplify the problem it can be assumed that the process has constant but unknown parameters. When the process is known, the design procedure specifies a set of desired control parameters. The adaptive controller should converge to these parameter values even when the process is unknown. A regulator with this property is called *self-tuning*, since it automatically tunes the controller to the desired performance.

The self-tuning regulator (STR) is based on the idea of separating the estimation of unknown parameters from the design of the controller. The basic idea is illustrated in Figure 2.4.

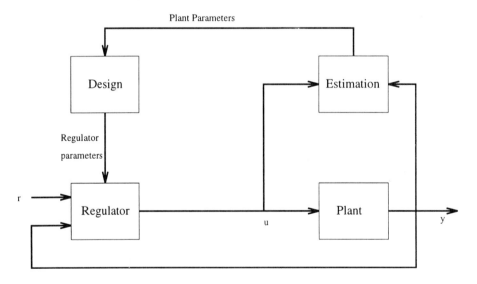

Figure 2.4: Block diagram of a self-tuning regulator.

The unknown parameters are estimated on-line, using a recursive estimation method. The estimated parameters are treated as if they were true. This is called the *certainty equivalence principle*. Many different estimation schemes

2.4. TRADITIONAL ADAPTIVE CONTROL METHODS

can be used, such as stochastic approximation, least squares, extended and generalized least squares, and maximum likelihood [13]. The block "Design" in Figure 2.4 represents an on-line solution to the design problem for a system with known parameters (*underlying design problem*). When the system parameters are unknown, the "design" is based on an estimation of the parameters coming from the block named "Estimation" (Figure 2.4). The design method is chosen depending on the specifications of the closed loop system. Different combinations of estimation methods and design methods lead to regulators with different properties.

As an example, consider the following discrete first-order system with unknown parameters a and b. The input-output relationship is

$$x(k+1) = a \cdot x(k) + b \cdot u(k) \qquad (2.2)$$

where $x(k)$ is the system output, $u(k)$ is the input and k is the time index. The control problem is to make the closed-loop system behave as if it were described by the equation

$$x(k+1) = c \cdot x(k) + d \cdot r(k) \qquad (2.3)$$

where the parameters c and d characterize the desired system response and $r(k)$ is the reference input (Figure 2.4). If the system parameters are known, the feedback controller should take the form

$$u(k) = \frac{c-a}{b} x(k) + \frac{d}{b} r(k) \qquad (2.4)$$

Since the parameters are assumed to be unknown, the least-square-error estimates will be used in place of the true values of a and b. The parameter estimates are derived from input-output measurements by thinking of the original system equations as a set of simultaneous equations in the parameters a and b. For example, if data is available at times $k = 0, 1, \ldots, n$ then writing out the system (2.2 and combining into matrix form yields

$$\begin{pmatrix} x(1) \\ x(2) \\ \vdots \end{pmatrix} = \begin{pmatrix} x(0) & u(0) \\ x(1) & u(1) \\ \vdots & \vdots \\ x(n-1) & u(n-1) \end{pmatrix} \cdot \begin{pmatrix} a \\ b \end{pmatrix} \qquad (2.5)$$

or more compactly

$$\mathbf{Y} = \mathbf{A} \cdot \mathbf{P} \qquad (2.6)$$

Least squares can then be applied to calculate the parameter estimates from the equation

$$\hat{\mathbf{P}} = \begin{pmatrix} \hat{a} \\ \hat{b} \end{pmatrix} = (\mathbf{A}^T \cdot \mathbf{A})^{-1} \cdot \mathbf{A}^T \cdot \mathbf{Y} \qquad (2.7)$$

where $\hat{a}$ and $\hat{b}$ are the estimates of the true parameters a and b used in the control law.

One of the advantages of the self-tuning regulator approach is that all the theoretical advances of least-square-error parameter identification can be applied to the learning part of the control system. That is, the theoretical lower bound on the parameter errors is known, and the propagation of the parameter error variance can be determined recursively. The disadvantages of this approach are that there is no guarantee of stability for the system and in fact, during the learning phase of the control, the input signal can be infinitely large and therefore not realizable. Another underlying problem is that the method requires an assumption of the structure of the system being controlled. That is, if the system being controlled does not match the assumed structure, the least-square-error parameter estimates may not completely describe the system and the closed loop system performance may not match the design specifications. In the simple case above, for example, the actual system being controlled must be completely described by a first order *linear* model in order to match the assumed structure. Also, the control system design is fixed in the sense that the design is intended to achieve a certain tracking bandwidth. If the reference input should change frequency, for example, the tracking performance will degrade and the design may need to be altered.

2.4.2 The Model-Reference Adaptive Control (MRAC)

The *model-reference adaptive control system* (MRAC) is one of the main approaches to adaptive control. The basic principle is illustrated in Figure 2.5.

The desired performance is expressed in terms of a reference model, which gives the desired response to a command signal. The system also has an ordinary feedback loop composed of the process and the regulator. The error e is

2.4. TRADITIONAL ADAPTIVE CONTROL METHODS

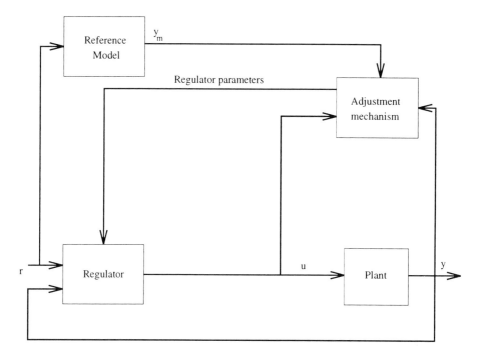

Figure 2.5: Block diagram of a model-reference adaptive system (MRAC).

the difference between the output of the system y and the reference model y_m. The regulator has parameters that are changed based on the error. There are thus two loops in Figure 2.5: an inner loop which provides the ordinary control feedback, and an outer loop which adjusts the parameters in the inner loop. The inner loop is assumed to be faster than the outer loop. Figure 2.5 is called a *parallel* MRAC. It is one of many possible ways of making a model-reference model.

There are essentially three basic approaches to the analysis and design of a MRAC:

- The gradient approach
- Lyapunov functions
- Passivity theory

The gradient method was used by Whitaker in the original work on the MRAC. This approach is based on the assumption that the parameters change more slowly than the other variables in the system. This assumption is essential for the computation of the sensitivity derivatives that are needed in the adaptation mechanism. The gradient approach will not necessarily result in a stable closed-loop system. This observation inspired the application of stability theory. Lyapunov's stability theory and passivity theory have been used to modify the adaptation mechanism. Important issues of the stability of model reference adaptive control methods, based on Lyapunov theory, are examined in [162].

Model-following is an important part of the MRAC, as in other adaptive controllers. A thorough discussion of model-following and regulator design is given in [13].

2.4.3 Limitations

The vast majority of the work on adaptive control has been on methods for linear or weakly nonlinear systems. This has some justification in that for small enough variations, any smooth system is linear; it also allows rigorous conclusions to be proven about controller behavior. Unfortunately, variations in the real world are not always small enough.

Linear adaptive control schemes have been widely studied and are widely used in industry. Nonlinear adaptive controls are less well developed and less well understood. The most common approach for the conventional methods is to use linear approximations of the plant to be controlled and some loosely defined response specifications such as bandwidth (speed of response) and phase-margins (degree of stability). The success of this approach depends heavily on the match between the actual system dynamics and the linearized approximation.

Nonlinear self-tuning controllers can offer substantially better performance than linear controllers, but the methods that have been used for nonlinear adaptive control are limited to processes with specific nonlinearities. Many controllers for nonlinear systems make the unrealistic assumption that exact nonlinear models of the process being controlled are known. Even those controllers which can include arbitrary nonlinear functions of the output and old inputs are typically limited to single input single output systems.

Speed is also a problem for real-time applications. The computational load is heavy for many conventional control methods and that makes them inappropriate for some real-world applications. Other problems of conventional control techniques appear when addressing physical effects such as friction, backlash, torque non-linearity (especially dead zone), high-frequency dynamics, and sen-

2.5 The OGY Method for the Control of Chaotic Systems

A method developed for "controlling chaos" is based on work by Ott, Grebogi and Yorke [156] (OGY). This method takes advantage of features specifically found in chaotic systems and unlike the methods described in the previous section cannot cope with changes in the dynamics of the system, i.e it is not applicable to adaptive control.

The first crucial development responsible for much of the recent progress associated with the OGY method was the realization that multi-dimensional phase portraits could be constructed from measurements of a *single* scalar time series, for example just one variable of the system. Here the portraits are constructed from the scalar time series using time delays as before, except that only a few (typically only one) scalar variables of the system are used. For an 'infinite amount' of data the delay step Δt can in principle be chosen almost arbitrarily [198]. However, experiments [173] show that the quality of the portraits depends on the value of Δt, and criteria for the selection of Δ were produced in [70].

A useful class of invariants which can be used to characterize a strange attractor is the set of all periodic trajectories. Since the strange attractor is the set of all points of the phase space visited by a trajectory after the transients have settled down, the trajectory from any point P on the strange attractor will make arbitrarily close returns to P. If the dynamics are smooth and nonlinear, by slightly varying some control parameter, one should be able to move P by a small amount so that the close return becomes exact, which implies that there is a periodic point arbitrarily close to P. This suggests that periodic points are dense on the strange attractor. Since the motion on the attractor is chaotic these cycles must be unstable.

The second critical observation is that: the structure of the strange attractor in the neighbourhood of a periodic point and the motion of points in this neighbourhood are determined by the tangent space of the periodic point. In particular, the eigenvalues give the local scaling observed in the strange attractor. Thus the nonlinear attractor can be considered as a collection of linear neighborhoods about the periodic points, i.e. the behavior of a chaotic system can be viewed as a collection of many orderly behaviors, none of which dominate under ordinary circumstances.

The OGY method applies these two observations to produce a viable method for controlling chaos. The first step in the process is to determine some of the unstable, low-period trajectories that are embedded in the chaotic attractor. For a particular case this is done by first selecting a suitable point

$$\xi_F(t) = (x(t), x(t - \Delta t), \ldots, x(t - p\Delta t)) \tag{2.8}$$

where $x(t)$ is one measurable scalar parameter of the dynamic system and $p \geq 1$ is a suitably chosen integer. We then monitor the dynamic system for a while and note all points which return closely to $\xi_F(t)$. For the purpose of simplicity, let us assume that in what follows this point is a fixed point of period 1, i.e. $p = 1$. In practice we would replace ξ_F by the mean of the near return points which we collected. Studying the short cycles will only allow us to determine the gross features of the dynamics, but does suffice to explain the underlying ideas of the OGY method.

Using this data a *local linearization* about ξ_F can be constructed using a least squares procedure to fit a $(p+1) \times (p+1)$ matrix which approximates the derivative (a first order approximation whose determinant is the Jacobian) of the return map. The locally linear behavior of the map in the vicinity of ξ_F is therefore approximately described by this matrix. Since the system is chaotic the absolute value of the largest eigenvalue of this matrix is larger than one and provides an estimate of the instability, or rate of separation, of the saddle orbit near ξ_F. Eigenvectors with the absolute value of eigenvalues less than one correspond to attractive or stable directions and are more difficult to estimate from experimental data because one cannot observe how different parts of the attractor, or associated return map, contract onto each other. However, knowledge of these eigenvalues derived from the experimentally determined matrix allows us, for example, to approximate the associated information dimension, and the eigenvectors give stable and unstable directions for the return map.

Let the dynamics of the system be represented by a $p + 1$ dimensional nonlinear map

$$\xi_{n+1} = f(\xi_n, c) \tag{2.9}$$

where the iteration $n \to n + 1$ corresponds to $t \to t + \Delta t$ and c is some scalar control parameter. Up to now c has been constant, say $c = c_0$. Finally, one

2.5. THE OGY METHOD

assumes that the position of the periodic orbit is a function of c, but that the local dynamics do not vary much with small changes in c about c_0.

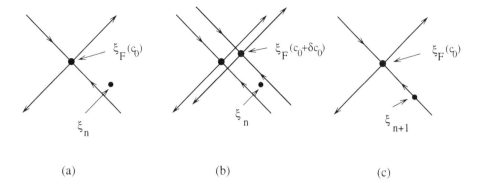

Figure 2.6: An example of the OGY method application.

The final step in the OGY method is a *sensitivity analysis*. By changing c slightly and observing how the desired orbit changes position, one can estimate the partial derivatives of the orbit location with respect to c. To control the chaotic system, one attempts to confine iterates of the map to a small neighbourhood of ξ_F. When an iterate falls near ξ_F we change c from its nominal value of c_0 by δc, as in Figure 2.6, thereby changing the location of the orbit and its stable directions, so that the next iterate will be forced back towards the stable manifold of the original orbit for $c = c_0$. Thus the OGY method rests on attempting to force the dynamics to stay in the neighbourhood of an unstable periodic orbit in the attractor and this makes it quite different from any other published methods for controlling chaos. These have mostly rested on developing a sufficiently detailed model to enable identification of the key parameters and then changing those parameters enough to take the system out of the chaotic regime.

The OGY method relies on small variations of one or more control parameters to stabilize the system into an (otherwise unstable) periodic motion natural to the system. For this reason it is termed a 'weak control' method.

Chapter 3

The Attitude Control Problem

There is no longer any need to argue that the communications satellite is ultimately going to have a profound effect upon society; the events of the last ten years have established this beyond question. Nevertheless, it is possible that even now we have only the faintest understanding of its ultimate impact upon our world.
Arthur C. Clarke
(UNESCO Space Communications Conference, Paris, 1969)

The orientation control of a rigid body has important applications from pointing and slewing of aircraft, helicopter, spacecraft and satellites, to the orientation control of a rigid object held by a single or multiple robot arms. A rigid body can be defined as a system of particles, in which the distances r_{ij} (r_{ij} is the distance between the ith and the jth particle) are fixed and cannot vary with time. In what follows, the rigid body will sometimes be referred to as satellite or spacecraft, since the need for attitude control systems arises frequently in aerospace technology, but the underlying definitions and methods will be generally applicable to rigid bodies.

The attitude of a satellite is its orientation in space. Its rigid body motion is specified by its position, velocity, attitude and attitude motion. The first two quantities describe the translational motion of the center of mass of the spacecraft and are the subject of what is variously called *celestial mechanics, orbit determination,* or *space navigation,* depending on the aspect of the problem that is emphasized. The latter two quantities describe the rotational motion of the body about the center of mass and are considered in this book.

Generally orbit and attitude are interdependent. For example, in a low altitude Earth orbit, the attitude will affect the atmospheric drag which will affect the orbit; the orbit determines the satellite position which determines both the atmospheric density and the magnetic field strength which will, in turn, affect the attitude. Although knowledge of the satellite orbit is frequently required to perform attitude determination and control functions, here this dynamical coupling will not be considered and it will be assumed that the time history of the rigid spacecraft position is known and has been supplied by some process external to the attitude determination and control system.

One distinction between orbit and attitude problems is related to their historical development. Predicting the orbital motion of celestial objects was the initial motivation of much of Newton's work. Thus, although the space age has brought with it vast new areas of orbit analysis, a large amount of theory directly related to celestial mechanics has existed for several centuries. In contrast, although some of the techniques are old, most of the advances in attitude determination and control have occurred since the launch of Sputnik I in 1957. The result of this is that even the language of attitude determination and control is still evolving.

Attitude analysis may be divided into determination, prediction and control. *Attitude determination* is the process of computing the orientation of a spacecraft relative to either an inertial reference or some object of interest, such as the Earth. This typically involves several types of sensors on each spacecraft and sophisticated data processing procedures. The accuracy limit is usually determined by a combination of processing procedures and system hardware. In the present book it will be assumed that the rigid body under consideration has sufficient sensors which are normally very accurate[1].

Attitude prediction is the process of forecasting the future orientation of the satellite using dynamical models. The limiting features are the knowledge of the applied and environmental torques and the accuracy of the mathematical model of satellite dynamics.

---

[1] This assumption does not hold in some of the simulation results of Chapter 8, where in order to test the robustness of the control system, artificial noise is introduced to the system sensors.

Attitude control is the process of orienting the satellite in a specified, predetermined direction. It consists of two areas: *attitude stabilization*, which is the process of maintaining an existing orientation, and *attitude maneuver control*, which is the process of controlling the reorientation of the satellite from one attitude to another. However, the two areas are not totally distinct. For example, the stabilization of a satellite with one axis towards the Earth implies a continuous change (maneuver) in its *inertial* orientation. The limiting factor for attitude control is typically the performance of the maneuver hardware and software, and the control electronics, although with autonomous control systems, it may be the accuracy of orbit or attitude information.

Some form of attitude determination and control is required for nearly all spacecraft. For engineering or flight-related functions, attitude determination is required only to provide a reference for control. Attitude control is required to avoid solar or atmospheric damage to sensitive components, to control heat dissipation, to point directional antennas and solar panels (for power generation), and to orient rockets used for orbit maneuvers.

A method for categorizing spacecrafts is the procedure by which they are stabilized. The simplest procedure is to spin the spacecraft. The angular momentum of a *spin-stabilized* spacecraft will remain approximately fixed in inertial space for extended periods, because the external torques which affect it are extremely small in most cases. However, the rotational orientation of the spacecraft about the spin axis is not controlled in such a system. If the orientation of three mutually perpendicular spacecraft axes must be controlled, then the spacecraft is *three-axis stabilized*. In this case, some form of active control is usually required because environmental torques, although small, will normally cause the spacecraft orientation to drift slowly. Three-axis stabilized spacecrafts may be either *non-spinning* (fixed in inertial space) or fixed relative to a possibly rotating reference frame, as occurs for an Earth satellite which maintains one face towards the Earth and therefore is spinning at one rotation per orbit. Many missions consist of some phases in which the spacecraft is spin-stabilized and some phases in which it is three-axis stabilized. Some spacecrafts have multiple components, some of which are spinning and some of which are three-axis stabilized.

In general, a spacecraft attitude control system consists of the following four major functional sections: sensing, logic, actuation, and vehicle dynamics. The logic programs the electronic signals in a correct sequence to the torque producing elements (actuation), which in turn rotate the spacecraft about its center of mass. The resulting motion (dynamics) is then monitored by the vehicle sensors which thus close the loop of the spacecraft attitude control system (Figure 3.1).

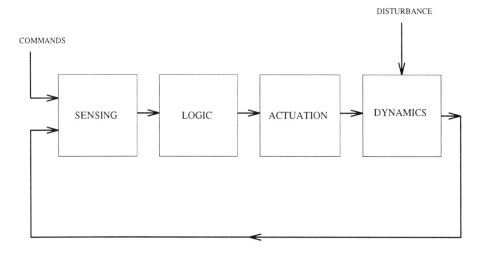

Figure 3.1: General satellite attitude control system

3.1 The Euler Equations

In the interest of completeness, the rigid-body equations are derived from first principles. Let δm, an element of mass of the object, have velocity $\mathbf{v}$ relative to inertial axes and let $\delta \mathbf{F}$ be the resultant force that acts upon it.

From Newton's second law

$$\delta \mathbf{F} = \delta m \frac{d\mathbf{v}}{dt} \tag{3.1}$$

The sum for all the elements δm gives

$$\sum \delta \mathbf{F} = \sum \delta m \frac{d\mathbf{v}}{dt} = \frac{d}{dt} \sum \mathbf{v} \delta m \tag{3.2}$$

The internal forces, that is, those exerted by one element upon another, all occur in equal and opposite pairs (Newton's third law) and hence contribute nothing to the summation. Thus $\sum \delta \mathbf{F} = \mathbf{F}$ is the resultant external force acting upon the object.

The velocity of δm is

$$\mathbf{v} = \mathbf{v}_c + \frac{d\mathbf{r}}{dt} \tag{3.3}$$

3.1. THE EULER EQUATIONS

where $\mathbf{v}_c$ is the velocity of the mass center, and $\mathbf{r}$ the distance of the element from the body mass center C. Thus

$$\sum \mathbf{v}\delta m = \sum \left(\mathbf{v}_c + \frac{d\mathbf{r}}{dt}\right)\delta m = m\mathbf{v}_c + \frac{d}{dt}\sum \mathbf{r}\delta m \qquad (3.4)$$

Since C is the mass center, $\sum \mathbf{r}\delta m = 0$, so that

$$\sum \mathbf{v}\delta m = m\mathbf{v}_c \qquad (3.5)$$

where m is the total mass of the object. Thus equation (3.1) becomes

$$\mathbf{F} = m\frac{d\mathbf{v}_c}{dt} \qquad (3.6)$$

This equation relates the external force on the object to the motion of the mass center. The relation between the external moment and the rotation of the object is also needed. The *moment of momentum* of δm is by definition $\delta \mathbf{h} = \mathbf{r} \times \mathbf{v}\delta m$. Consider

$$\frac{d}{dt}(\delta \mathbf{h}) = \frac{d}{dt}(\mathbf{r} \times \mathbf{v})\delta m = \frac{d\mathbf{r}}{dt} \times \mathbf{v}\delta m + \mathbf{r} \times \frac{d\mathbf{v}}{dt}\delta m \qquad (3.7)$$

Now from equation (3.3)

$$\frac{d\mathbf{r}}{dt} = \mathbf{v} - \mathbf{v}_c \qquad (3.8)$$

Also from equation (3.1),

$$\mathbf{r} \times \frac{d\mathbf{v}}{dt}\delta m = \mathbf{r} \times \delta \mathbf{F} = \delta \mathbf{G} \qquad (3.9)$$

where $\delta \mathbf{G}$ is the moment of $\delta \mathbf{F}$ about C.

Equation (3.7) then becomes

$$\delta \mathbf{G} = \frac{d}{dt}(\delta \mathbf{h}) - (\mathbf{v} - \mathbf{v}_c) \times \mathbf{v}\delta m = \frac{d}{dt}(\delta \mathbf{h}) + \mathbf{v}_c \times \mathbf{v}\delta m \qquad (3.10)$$

because $\mathbf{v} \times \mathbf{v} = \mathbf{0}$. Summing up now for all the elements yields

$$\sum \delta \mathbf{G} = \frac{d}{dt}\sum \delta \mathbf{h} + \mathbf{v}_c \times \sum \mathbf{v}\delta m \qquad (3.11)$$

$\sum \delta \mathbf{h}$ is the *angular momentum* of the object and is denoted by $\mathbf{h}$. Using equation (3.5) and noting that $\mathbf{v}_c \times \mathbf{v}_c = 0$ equation (3.11) reduces to

$$\mathbf{G} = \frac{d\mathbf{h}}{dt} \qquad (3.12)$$

Note that both **G** and **h** are referred to a moving point, the mass center. For a moving reference point other than the mass center the equation does not apply in general.

So finally the two vector equations that describe the motion of a rigid body are

$$\mathbf{F} = m\frac{d\mathbf{v}_c}{dt} \tag{3.13}$$

$$\mathbf{G} = \frac{d\mathbf{h}}{dt} \tag{3.14}$$

Let the *angular velocity* of the object be

$$\boldsymbol{\omega} = \mathbf{i}\omega_1 + \mathbf{j}\omega_2 + \mathbf{k}\omega_3$$

where $\mathbf{i}, \mathbf{j}, \mathbf{k}$ are unit vectors in the directions of x, y, z.

Now the velocity of a point in a rotating rigid body is given by

$$\mathbf{v} = \mathbf{v}_c + \boldsymbol{\omega} \times \mathbf{r} \tag{3.15}$$

thus

$$\mathbf{h} = \sum \mathbf{r} \times \mathbf{v}_c \delta m + \sum \mathbf{r} \times (\boldsymbol{\omega} \times \mathbf{r})\delta m \tag{3.16}$$

Since $\mathbf{v}_c$ is constant with respect to the summation, and since $\sum \mathbf{r}\delta m = 0$, then the first sum in the previous equation is zero.

The second sum is easily expanded by the rule for a vector triple product.

$$\mathbf{r} \times (\boldsymbol{\omega} \times \mathbf{r}) = \boldsymbol{\omega}(\mathbf{r} \cdot \mathbf{r}) - \mathbf{r}(\boldsymbol{\omega} \cdot \mathbf{r}) = \boldsymbol{\omega}\mathbf{r}^2 - \mathbf{r}(\boldsymbol{\omega} \cdot \mathbf{r})$$

Since the vector **r** has components (x, y, z) we have

$$\mathbf{h} = \boldsymbol{\omega}\sum(x^2 + y^2 + z^2)\delta m - \sum \mathbf{r}(\omega_1 x + \omega_2 y + \omega_3 z)\delta m \tag{3.17}$$

The scalar components of the previous equation are

$$h_x = \omega_1 \sum(y^2 + z^2)\delta m - \omega_2 \sum xy\delta m - \omega_3 \sum xz\delta m$$

$$h_y = -\omega_1 \sum xy\delta m + \omega_2 \sum(x^2 + z^2)\delta m - \omega_3 \sum yz\delta m$$

$$h_z = -\omega_1 \sum xz\delta m - \omega_2 \sum yz\delta m + \omega_3 \sum(x^2 + y^2)\delta m$$

3.1. THE EULER EQUATIONS

The summations that occur in these equations are the moments and products of inertia of the rigid body, and substituting symbols for these sums we get

$$\begin{aligned} h_x &= I_x\omega_1 - I_{xy}\omega_2 - I_{xz}\omega_3 \\ h_y &= -I_{xy}\omega_1 + I_y\omega_2 - I_{yz}\omega_3 \\ h_z &= -I_{xz}\omega_1 - I_{yz}\omega_2 + I_z\omega_3 \end{aligned} \quad (3.18)$$

However, it is possible to find a set of Cartesian axes for which the inertia matrix will be diagonal (see [195]). The axes are called the *principal axes* and the corresponding diagonal elements I_x, I_y, I_z are known as the *principal moments of inertia*. Then the above equations take the form

$$\begin{aligned} h_x &= I_x\omega_1 \\ h_y &= I_y\omega_2 \\ h_z &= I_z\omega_3 \end{aligned} \quad (3.19)$$

As can be seen from equation (3.14), the derivatives of the above equations with respect to time are required. If the axes of reference are non-rotating, then the equations will have a complicated form. So, in the first instance it is desirable to have the frame of reference $Cxyz$ fixed to the object.

The derivative of a vector $\mathbf{A}$, referred to a frame of reference rotating with angular velocity $\boldsymbol{\omega}$, is

$$\frac{d\mathbf{A}}{dt} = \frac{\partial \mathbf{A}}{\partial t} + \boldsymbol{\omega} \times \mathbf{A} \quad (3.20)$$

where

$$\frac{\partial \mathbf{A}}{\partial t} = \mathbf{i}\frac{dA_x}{dt} + \mathbf{j}\frac{dA_y}{dt} + \mathbf{k}\frac{dA_z}{dt}$$

The vector equations of motion (3.13) and (3.14) then become, when referred to the frame of reference $Cxyz$, fixed to the rigid body,

$$\mathbf{F} = m\frac{\partial \mathbf{v}_c}{\partial t} + m\boldsymbol{\omega} \times \mathbf{v}_c \quad (3.21)$$

$$\mathbf{G} = \frac{\partial \mathbf{h}}{\partial t} + \boldsymbol{\omega} \times \mathbf{h} \quad (3.22)$$

These equations have the scalar components

$$\begin{aligned} F_x &= m(\dot{U} + \omega_2 W - \omega_3 V) \\ F_y &= m(\dot{V} + \omega_3 U - \omega_1 W) \\ F_z &= m(\dot{W} + \omega_1 V - \omega_2 U) \end{aligned} \quad (3.23)$$

and

$$\begin{align}
L &= I_x\dot{\omega}_1 - (I_y - I_z)\omega_2\omega_3 \\
M &= I_y\dot{\omega}_2 - (I_z - I_x)\omega_3\omega_1 \\
N &= I_z\dot{\omega}_3 - (I_y - I_z)\omega_2\omega_3
\end{align} \quad (3.24)$$

where $\mathbf{v}_c = (U, V, W)$ and $\mathbf{G} = (L, M, N)$.

Equations (3.23) and (3.24) are the *Euler equations* of motion of the rigid body. These equations are identical to the equations for the bending of a thin elastic rod or tape. This analogy was discovered by Kirchoff in 1831.

As noted earlier, in the model considered in most of this book it is assumed that there are no external forces, so from now when referring to the Euler equations, it is assumed that equations (3.24) are the Euler equations.

3.2 Chaos in the Euler Equations

There are several cases where freely rotating rigid bodies can exhibit chaotic behavior [126], [145]. The first case is when one of the moments (L, M or N) varies periodically in time. The second case is where one has parametric excitation through time-periodic changes in the principal inertias, for example, $I_y = I_0 + B\cos\Omega t$.

The third case comes from the schemes based on linear or quadratic feedback, which have been proposed to stabilize rigid body spacecraft attitude, described by the Euler equations (3.24). A simple scheme of this type employs jets to impart torque according to suitable linear combinations of the sensed angular velocities $\boldsymbol{\omega}$ about the body-fixed principal axes[2].

If the torque feedback matrix for the nonlinear system is denoted by $\mathbf{A}$, so that $\mathbf{G} = (L, M, N) = \mathbf{A}\boldsymbol{\omega}$, the equations (3.24) become

$$\begin{align}
(\mathbf{A}\boldsymbol{\omega})_1 &= I_x\dot{\omega}_1 - (I_y - I_z)\omega_2\omega_3 \\
(\mathbf{A}\boldsymbol{\omega})_2 &= I_y\dot{\omega}_2 - (I_z - I_x)\omega_3\omega_1 \\
(\mathbf{A}\boldsymbol{\omega})_3 &= I_z\dot{\omega}_3 - (I_y - I_z)\omega_2\omega_3
\end{align} \quad (3.25)$$

It has been noticed [126], [164] that for certain choices of I_x, I_y, I_z and $\mathbf{A}$ equations (3.25) exhibit both strange attractors and limit cycles. Since these linear feedback rigid body motion (LFRBM) equations are slightly more complicated

---
[2]R.B Leipnik derived similar equations for the psychology of multiple conflicting objectives, conflict with intervention, etc. [126].

than Lorenz's equations (2.1) (see [126]), this conclusion is not surprising. The interpretation of strange attractor motion, in this case, is that the body executes a wobbly spin first about one, then about the other of a conjugate pair of directions fixed relative to the body axes (and symmetric about an axis). The limiting spin magnitudes and directions define rest points of the system, called eye-attractors because of their appearance.

Figures 3.2, 3.3, 3.4, show the double strange attractor of equations (3.25), in the xy, xz and yz planes respectively, for $I_x = 3, I_y = 2, I_z = 1$ and

$$\mathbf{A} = \begin{pmatrix} -1.2 & 0 & \sqrt{6}/2 \\ 0 & 0.35 & 0 \\ -\sqrt{6} & 0 & -0.4 \end{pmatrix} \tag{3.26}$$

The attractor of an orbit is determined by the location of the initial point of that orbit.

3.3 Orientation of a Rigid Body

Since the frame of reference adopted for the equations of motion is fixed to the rigid body, and moves with it, the position and orientation of the object cannot be described relative to this frame.

For this purpose an inertial frame of reference $Ox'y'z'$ is introduced (see Figure 3.5). Let Oz' be taken vertically downward, and Ox' horizontal in the vertical plane containing the initial velocity vector of the mass center. The origin O is assumed to coincide with C at $t = 0$.

The orientation of the rigid body is then given by a series of three consecutive rotations, whose order is important. The object is imagined first to be oriented so that its axes are parallel to $Ox'y'z'$. It is then in the position $Cx_1y_1z_1$. The following rotations are then applied.

1. A rotation Ψ about Oz_1, carrying the axes to $Cx_2y_2z_2$ (bringing Cx to its final azimuth).

2. A rotation Θ about Oy_2, carrying the axes to $Cx_3y_3z_3$ (bringing Cx to its final elevation).

3. A rotation Φ about Ox_3, carrying the axes to their final position $Cxyz$ (giving the final angle of bank to the wings, if we imagine that our rigid body is an airplane).

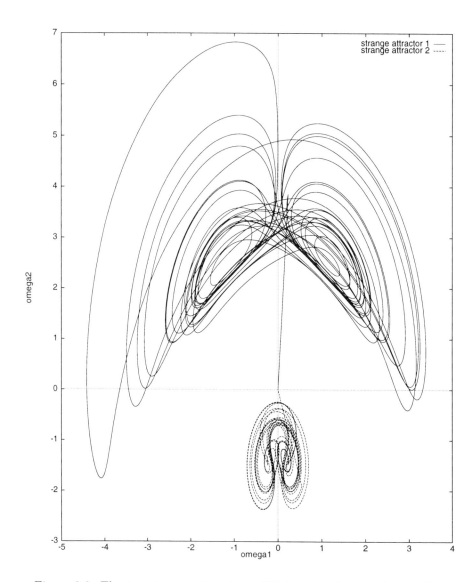

Figure 3.2: The two strange attractors of Euler equations in the xy plane.

3.3. ORIENTATION OF A RIGID BODY

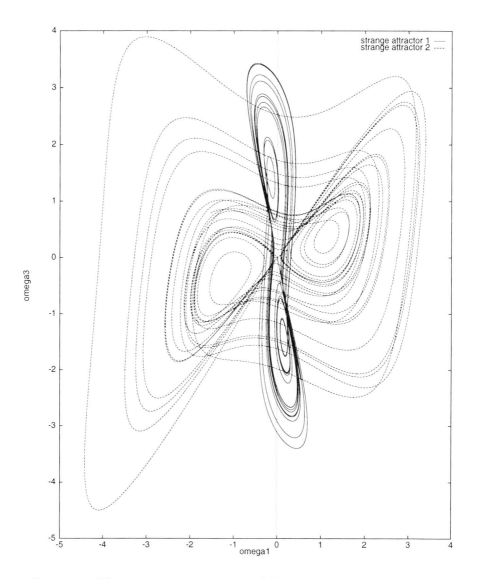

Figure 3.3: The two strange attractors of Euler equations in the xz plane.

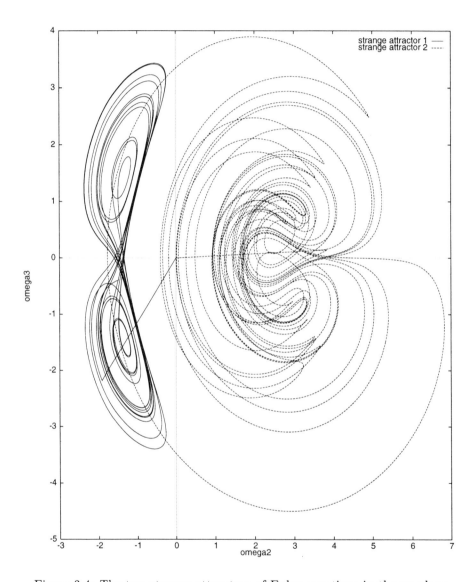

Figure 3.4: The two strange attractors of Euler equations in the yz plane.

3.3. ORIENTATION OF A RIGID BODY

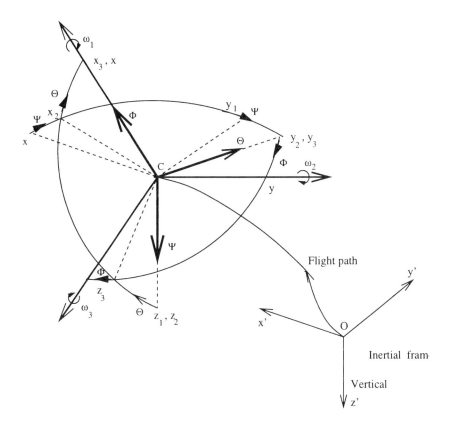

Figure 3.5: Rigid body orientation.

Now we wish to express the orientation of the rigid body in terms of the angular velocity components $(\omega_1, \omega_2, \omega_3)$. Let $(\mathbf{i}, \mathbf{j}, \mathbf{k})$ be unit vectors, with subscripts 1,2,3 denoting directions (x_1, y_1, z_1), etc.

Let the object experience, in time Δt, an infinitesimal rotation from the position defined by Φ, Θ, Ψ to that corresponding to $(\Phi + \Delta\Phi), (\Theta + \Delta\Theta), (\Psi + \Delta\Psi)$. The vector representing this rotation is approximately

$$\Delta \mathbf{n} \doteq \mathbf{i}\Delta\Phi + \mathbf{j}_3\Delta\Theta + \mathbf{k}_2\Delta\Psi \tag{3.27}$$

and the angular velocity is exactly

$$\boldsymbol{\omega} = \lim_{\Delta t \to 0} \frac{\Delta \mathbf{n}}{\Delta t} = \mathbf{i}\dot\Phi + \mathbf{j}_3\dot\Theta + \mathbf{k}_2\dot\Psi \tag{3.28}$$

If the components of $\boldsymbol{\omega}$ given in equation (3.28) are projected onto $Cxyz$, and remembering that $\boldsymbol{\omega} = \mathbf{i}\omega_1 + \mathbf{j}\omega_2 + \mathbf{k}\omega_3$, the necessary relations are obtained. There is another way to derive these relations. A rotation of the axes that are fixed to the body is made, to a system of axes which is fixed to the mass center but with steady inertial orientation this time. The rotation is again described by the steps 1, 2, 3 above.

So, the vector $(0, 0, \dot\Psi)$ is rotated about y_1 axes by an angle Θ:

$$(0, 0, \dot\Psi) \cdot \begin{pmatrix} \cos\Theta & 0 & \sin\Theta \\ 0 & 1 & 0 \\ -\sin\Theta & 0 & \cos\Theta \end{pmatrix} = (-\dot\Psi \sin\Theta, 0, \dot\Psi \cos\Theta) \tag{3.29}$$

Now, the vector $(-\dot\Psi \sin\Theta, \dot\Theta, \dot\Psi \cos\Theta)$ is rotated about x_1 axis by an angle Φ:

$$(-\dot\Psi \sin\Theta, \dot\Theta, \dot\Psi \cos\Theta) \cdot \begin{pmatrix} 1 & 0 & 0 \\ 0 & \cos\Phi & -\sin\Phi \\ 0 & \sin\Phi & \cos\Phi \end{pmatrix} =$$

$$(-\dot\Psi \sin\Theta, \dot\Theta \cos\Phi + \dot\Psi \sin\Phi \cos\Theta, -\dot\Theta \sin\Phi + \dot\Psi \cos\Theta \cos\Phi) \tag{3.30}$$

and this finally yields:

$$\begin{aligned} \omega_1 &= \dot\Phi - \dot\Psi \sin\Theta \\ \omega_2 &= \dot\Theta \cos\Phi + \dot\Psi \cos\Theta \sin\Phi \\ \omega_3 &= \dot\Psi \cos\Theta \cos\Phi - \dot\Theta \sin\Phi \end{aligned} \tag{3.31}$$

3.4. DEFINING THE PROBLEM

In order to find the angles (Φ, Θ, Ψ), in terms of $(\omega_1, \omega_2, \omega_3)$, the above equations must be solved. If they are regarded as algebraic equations in $\dot{\Phi}, \dot{\Theta}, \dot{\Psi}$, we obtain

$$\begin{aligned}\dot{\Theta} &= \omega_2 \cos\Phi - \omega_3 \sin\Phi \\ \dot{\Phi} &= \omega_1 + \omega_2 \sin\Phi \tan\Theta + \omega_3 \cos\Phi \tan\Theta \\ \dot{\Psi} &= (\omega_2 \sin\Phi + \omega_3 \cos\Phi) \sec\Theta \end{aligned} \qquad (3.32)$$

This system can be written:

$$\begin{pmatrix}\dot{\Theta} \\ \dot{\Phi} \\ \dot{\Psi}\end{pmatrix} = \begin{pmatrix} 0 & \cos\Phi & -\sin\Phi \\ 1 & \sin\Phi \tan\Theta & \cos\Phi \tan\Theta \\ 0 & \sin\Phi \sec\Theta & \cos\Phi \sec\Theta \end{pmatrix} \cdot \begin{pmatrix}\omega_1 \\ \omega_2 \\ \omega_3\end{pmatrix} \qquad (3.33)$$

This is a system of nonlinear differential equations and has to be integrated in conjunction with (3.24) to obtain the orientation angles Φ, Θ, Ψ.

3.4 Defining the Problem

The desire to expand permanent human presence within and beyond our solar system will probably necessitate the design and deployment of larger, potentially more complex spacecraft than any previously flown. During their anticipated long lifetime, these spacecraft will undergo significant configuration changes, resulting from initial and evolutionary on-orbit assembly, and from routine operations such as docking and berthing of interplanetary transportation vehicles.

Generally, in the operation of space vehicles, there is always a finite probability that a malfunction will occur, which results in uncontrolled tumbling of a vehicle. In a manned vehicle such uncontrolled motion creates a hazardous environment for the crew, who would experience oscillating accelerations. The structural integrity of the disabled vehicle may be jeopardized by prolonged tumbling, presenting additional danger. This book addresses the problem of detumbling a space station or satellite (reducing its kinetic energy) while achieving a desired orientation.

It is assumed that the satellite is a rigid body. Its rotational motion is described by the Euler equations (3.24), and its orientation in space is defined in terms of equations (3.32). The system is equipped with reaction thrusters, which provide control torques about the three principal body axes. Most of the time, it is assumed that no external forces (besides the control torques) are acting upon the system. The above system is highly nonlinear, and exhibits chaotic behavior under certain circumstances (section 3.2).

In addition, the dynamic system may change its characteristics. This could happen due to damage, malfunction, or a change in the principal moments of inertia. In this event, the available controller is no longer appropriate for the actual dynamic system. Thus the problem becomes adaptive, and an adaptive controller is required for the detumbling and orientation of the satellite.

3.5 Attitude Control Approaches

Most of the work on the attitude control problem was done after 1957. Many authors have tried to derive new control rules for the problem. Although quite old (almost 30 years ago), the work of Meyer in NASA [139], [140] is still considered today, one of the most seminal investigations of the attitude control problem. Meyer, after expressing the attitude error in terms of an error matrix, synthesized a class of control laws for which the control inputs (torques) are functions of the real eigenvector of the error matrix [42], [75], [225] and the angular velocity of the controlled body. However, the method is not applicable to on-off control[3], since the results are valid mostly for reaction wheel control systems. Besides that, the method is completely inapplicable to the *adaptive* attitude control problem (when the system dynamics change in an unknown way), since the control laws derived assume that the analytic form of the dynamic system under consideration is known.

In [104], Jahangir and Howe address the problem of minimum-time attitude control of a symmetric rigid body missile. They describe a scheme of generating thruster firing times as functions of the initial and desired state of the missile. The scheme involves the transformation of the state variables and the integration of the transformed states and equations. Although they avoid the iterative part of solving a two boundary value problem, their method assumes memory data storage for table lookup, and numerical integration.

Singh et al. [188] used linearization theory to represent the nonlinear dynamics of a space station and discussed the attitude control problem for space vehicles employing control moment gyros. A similar approach using also linearization theory is considered in [89], but this time the problem of a spinning

---

[3]For example, reaction jet control systems.

3.5. ATTITUDE CONTROL APPROACHES

satellite with two small jets is discussed. Attitude control using eigenvector analysis on the linearized attitude equations of motion, for a spinning symmetrical satellite in an elliptic orbit, is used in [111]. Attitude control of an inverted pendulum using linearized dynamics of the system is discussed in [71]. [118] presents a linear regulator feedback law for controlling a tethered satellite. The dynamics and the design considerations for the attitude control of a two-dimensional tethered satellite can be found in [102], [127]. Paielli and Bach [158] used linear error dynamics (by considering only the linear terms in the Euler parameters) to derive an attitude control law.

[130] introduces a coordinate frame for the rest-to-rest reorientation of a satellite, which transforms the original nonlinear problem into a linear one. The approach is especially attractive for optimal control attitude control, since it reduces the 7×7 matrix computations to 2×2 matrix computations. In [154] a nonlinear *observer* is proposed, for reconstructing the state variables of a spacecraft. Then existing feedback control laws are used, giving a system which is asymptotically stable in a specific region. An observer based method, for a reaction wheel attitude controller, is proposed in [200], while various control laws for a three reaction wheel, three magnetic torque configuration are found in [112].

[21] discusses methods of geometric mechanics for stability analysis of rigid body dynamics. Another feedback controller incorporating gain-scheduled adaptation of the attitude gains is developed for a linearized model of a gravity gradient stabilized spacecraft in [163]. The simulations are done for the Space Station Freedom. Wie et al. evaluate the performance and stability of both classical and modern controllers for the Space Station Freedom [224]. A simulator description of the Space Station Freedom with the McDonnell Douglas Space Systems Co./Honeywell Attitude Control System can be found in [165]. More general methods for the design of spacecraft simulators can be found in [202]. A plume flow model to calculate the forces and the heat transfer caused by the firing of attitude control thrusters on satellites is developed in [54].

In [164], it is shown, based on an analysis of a simple-axis problem, that if the momentum wheel assembly performance parameters are adequately matched to the spacecraft, it is possible to achieve a globally stable equilibrium. S. Vadali [204] derives various attitude control laws, for a space station, in the case of absence of disturbances using Lyapunov's second method. Satellite attitude control laws using Lyapunov's methods are also the subjects of [23], [39]. The application of a game theoretic control approach, combined with an internal feedback loop decomposition for uncertainties in the moments of inertia of a space station (which are considered constant in time) is described in [171].

[189] considers a control law, for a class of uncertain nonlinear systems which can be decoupled by state variable feedback. The law is based on the technique of variable structure and is applied for the control of an orbiting spacecraft which uses reaction jets. Uncertainties in the spacecraft parameters are also considered in [183], [184]. In [51] a method based on the algebraic theory of rational fractions, for the control of a spinning satellite using gyro-torquers, is discussed. Attitude control using gyro-torquers is also considered in [103]. The attitude control (pitch and roll) of automobiles is the topic in [133].

Optimal control using nonlinear programming techniques applied to satellite attitude control is discussed in [87]. Gregory and Peters propose a device [81] for the direct measurement of the attitude stability of a space station, while [180] present a technique for attitude determination. Three axis attitude control of a rigid body spacecraft using a sliding-mode control law is described in [57]. The approach is valid as long as sliding motion is maintained and the extreme values of the plant dynamic parameters are known.

In [161] a Kalman filter is used to estimate local magnetic field and perform magnetic attitude control for the GP-B satellite. Davis and Levinson [50] introduced the use of a gravity-stabilized tether attached to a nonspinning part of a satellite, for enabling its attitude control by employing conventional control systems.

An enhancement in the solving techniques for the two-point boundary value optimal attitude control problem is presented in [129]. Various problems, advantages and disadvantages of different choices, associated with the design of a space station control system, are discussed in [26]. [185], [186] present an approach to the three-axis attitude control of a spacecraft-beam-tip body system, based on invertibility results. Momentum management systems designed to face the problem of momentum saturation of the control moment gyros, due to noncyclical external torques acting on the space station, are considered in [228].

A near optimal orbit and attitude control system, for a plate-like rigid spacecraft in geostationary orbit, is presented in [169]. All the results and conclusions are based on simple linear models. A fuzzy logic controller for the control of a spacecraft is applied in [46]. [182] proposes a control law for single axis rotational maneuvers of a spacecraft-beam-tip body (an antenna

3.5. ATTITUDE CONTROL APPROACHES

or reflector), in the presence of an unknown but bounded disturbance torque, acting on the spacecraft. In [190], impulse response functions are used for the selection of control switch times, in the bang-bang control of linear, elastic slewing satellites. A bang-bang control law is also presented in [56] and it is further extended in [58]. The nullification of the accumulated effect of the modeling errors, achieved by a correction in the control to induce the physical system to have a behavior close to the reference model, is the subject of [38]. A parameter optimization procedure is applied to find the gains of the described method.

Yuan and Stieber [229] employed a PI compensator, augmented by a Kalman filter, to control the communications beams and the attitude angles of a flexible spacecraft. They explored two design methods: the first one based on eigenvalue analysis and the second based on singular value criteria. A review in attitude control systems and beam pointing accuracy can be found in [146], while a general framework for the analysis of attitude tracking control problems can be found in [208]. [187] proposes a control law for asymptotic function reproducibility of a class of nonlinear systems, such that the output of the system tends asymptotically to a given function. Based on this control law, a nonlinear feedback control law is then derived for the attitude control of a satellite containing symmetric rotors.

The application of a controller consisted of a servo-compensator, a stabilizing feedback loop and a feedforward compensator, to the design of a vertical takeoff and landing aircraft flight control system is discussed in [136]. Hess [91] studied aircraft attitude control systems, based on the optimal control model of the human pilot. The optimal control human pilot model has its genesis in the hypothesis that, with limitations and in specific well-defined control tasks, the human pilot can be described in terms of the operation of a linear optimal estimator and regulator. Geometric control theory for rigid body attitude control is considered in [44]. In addition an algorithm for stabilizing the system is outlined as proposed by [88]. A simple two-surface solar controller is described and applied for the attitude control of a spacecraft in [159], [206]. A proportional plus derivative control law for attitude control of non rigid body spacecraft is found in [69]. In [179] a decomposed controller, which consists of two coupled electronic integrators is introduced, for satellite control.

The dynamics of UOSAT (low orbit satellite with a principal axis pointing towards the Earth center and a minimum number of sensors and hardware) and its control using the on-board magnetorquer is given in [96]. [41] describes the dynamic modeling and control of the SPOT French Earth observation satellites. In [105] it is shown that the knowledge of the Voyager's limit cycle motion, as measured by the celestial and the inertial sensors, is adequate to estimate a selected number of errors, which adversely affect the spacecraft attitude knowledge.

To summarize, so far none of the approaches to the attitude control problem have tried to use neural networks and genetic algorithms, techniques that have recently shown encouraging results in the control of simple systems [141]. In addition, most of the attitude control techniques use linearized equations of rigid body motion, something which is not valid over a large part of the dynamic system phase space.

Artificial neural networks and genetic algorithms are the main tools adopted in this book for addressing the attitude control problem. Rather than using linearized equations of motion, the highly nonlinear equations describing the attitude control problem are directly used in the book approach. Finally *hitherto*, other approaches to the attitude control problem have not dealt with the adaptive case (when the system dynamics change dramatically)[4]. In contrast, one of the main issues in this book is the adaptive attitude control problem.

3.6 Numerical Methods for Solving Ordinary Differential Equations

In many engineering and scientific problems, differential equations are in a form that it is not possible to solve analytically. It is therefore natural to consider numerical procedures for obtaining an approximate solution of the problem.

For example, consider the first order differential equation

$$y' = f(x, y) \qquad (3.34)$$

and the initial condition

$$y(x_0) = y(0) \qquad (3.35)$$

A numerical method for solving this initial value problem is a procedure for constructing approximate values $y_0, y_1, y_2, \ldots, y_n, \ldots$ of the solution $\phi(x)$ at the points $x_0 < x_1 < x_2 \cdots < x_n < \cdots$. More precisely, such a numerical procedure

---

[4]Actually a few approaches address the attitude control problem with only small variations in the system parameters.

3.6. NUMERICAL METHODS FOR SOLVING ODE

is referred to as a *discrete variable method* since the original problem involving continuous variables is replaced by one involving discrete variables.

In the first instance y_1 is determined, knowing y_0 from the initial condition and $\phi'(x_0) = f(x_0, y_0)$ from the differential equation (3.34). Then given y_1, y_2 is computed and so on. In determining y_2 one can either use the same method as in going from x_0 to x_1, or can use a different method that takes account of the fact, that both y_0 at x_0 and y_1 at x_1 are now known.

Methods that require only a knowledge of y_n, to determine y_{n+1}, are known as *one step* or *starting methods*. Methods that make use of data at more than the previous point, say y_n, y_{n-1}, y_{n-2} to determine y_{n+1} are known as *multi-step* or *continuing methods*.

There are several issues to consider when applying a numerical method. First, there is the question of *convergence*. That is, as the spacing between the points $x_0, x_1, x_2, \ldots, x_n, \ldots$ approaches zero, do the values of the numerical solution $y_1, y_2, \ldots, y_n, \ldots$ approach the values of the exact solution? There is also the question of estimating the error made in computing the values $y_1, y_2, \ldots, y_n$. This error arises from two sources: first, the formula used in the numerical method is only an approximate one, which causes a *discretization error*, or *truncation error*; second, due to the hardware it is possible to carry only a limited number of digits in any computation, which causes a *round-off error*.

3.6.1 The Runge-Kutta Method

The Runge-Kutta formula [25], [92], [123] is equivalent to a five-term Taylor formula

$$y_{n+1} = y_n + hy'_n + \frac{h^2}{2!}y''_n + \frac{h^3}{3!}y'''_n + \frac{h^4}{4!}y_n^{iv}$$

where $h = \Delta x = x_{n+1} - x_n$.

The Runge-Kutta formula involves a weighted average of values of $f(x, y)$ taken at different points in the interval $x_n \leq x \leq x_{n+1}$.

Consider the system of differential equations

$$\begin{aligned} x' = \frac{dx}{dt} &= f(t, x, y, z) \\ y' = \frac{dy}{dt} &= g(t, x, y, z) \\ z' = \frac{dz}{dt} &= v(t, x, y, z) \end{aligned} \quad (3.36)$$

with $x(t_0) = x_0, y(t_0) = y_0, z(t_0) = z_0$.

Then the Runge-Kutta formula is given by

$$\begin{aligned} x_{n+1} &= x_n + \frac{1}{6}h(k_{n1} + 2k_{n2} + 2k_{n3} + k_{n4}) \\ y_{n+1} &= y_n + \frac{1}{6}h(l_{n1} + 2l_{n2} + 2l_{n3} + l_{n4}) \\ z_{n+1} &= z_n + \frac{1}{6}h(m_{n1} + 2m_{n2} + 2m_{n3} + m_{n4}) \end{aligned} \qquad (3.37)$$

where

$$\begin{aligned} k_{n1} &= f(t_n, x_n, y_n, z_n) \\ k_{n2} &= f(t_n + \tfrac{1}{2}h, x_n + \tfrac{1}{2}hk_{n1}, y_n + \tfrac{1}{2}hl_{n1}, z_n + \tfrac{1}{2}hm_{n1}) \\ k_{n3} &= f(t_n + \tfrac{1}{2}h, x_n + \tfrac{1}{2}hk_{n2}, y_n + \tfrac{1}{2}hl_{n2}, z_n + \tfrac{1}{2}hm_{n2}) \\ k_{n4} &= f(t_n + h, x_n + hk_{n3}, y_n + hl_{n3}, z_n + hm_{n3}) \\ l_{n1} &= g(t_n, x_n, y_n, z_n) \\ l_{n2} &= g(t_n + \tfrac{1}{2}h, x_n + \tfrac{1}{2}hk_{n1}, y_n + \tfrac{1}{2}hl_{n1}, z_n + \tfrac{1}{2}hm_{n1}) \\ l_{n3} &= g(t_n + \tfrac{1}{2}h, x_n + \tfrac{1}{2}hk_{n2}, y_n + \tfrac{1}{2}hl_{n2}, z_n + \tfrac{1}{2}hm_{n2}) \\ l_{n4} &= g(t_n + h, x_n + hk_{n3}, y_n + hl_{n3}, z_n + hm_{n3}) \\ m_{n1} &= v(t_n, x_n, y_n, z_n) \\ m_{n2} &= v(t_n + \tfrac{1}{2}h, x_n + \tfrac{1}{2}hk_{n1}, y_n + \tfrac{1}{2}hl_{n1}, z_n + \tfrac{1}{2}hm_{n1}) \\ m_{n3} &= v(t_n + \tfrac{1}{2}h, x_n + \tfrac{1}{2}hk_{n2}, y_n + \tfrac{1}{2}hl_{n2}, z_n + \tfrac{1}{2}hm_{n2}) \\ m_{n4} &= v(t_n + h, x_n + hk_{n3}, y_n + hl_{n3}, z_n + hm_{n3}) \end{aligned}$$

There is no reason why the step size h needs to be kept fixed over the entire interval. It is possible to vary the step size according to some error criterion [42]. However the use of variable step sizes adds considerably to the complexity of an algorithm and leads to results at a set of nonuniformly spaced points which may not be useful in some applications. Halving and doubling intervals is generally more acceptable. On the other hand, algorithms with automatic step size control provide very good estimates of accuracy, and are overall quite efficient.

The local discretization error, in using the Runge-Kutta described above, is proportional to h^5 and as can be seen, the method requires a lot of computation. For this reason an improved method is examined in the next section.

3.6.2 The Adams-Moulton Method

In this section we consider a multi-step method, the Adams-Moulton predictor-corrector method [42], [116]. Before the method can be used, it is necessary to compute y_1, y_2, y_3 by some starting method, such as the Runge-Kutta method(the starting method should be as accurate as the multi-step method).

The formulas of this method for the solution of the system (3.36) are

$$\begin{aligned}x_{n+1} &= x_n + \frac{1}{24}h(55x'_n - 59x'_{n-1} + 37x'_{n-2} - 9x'_{n-3}) \\ y_{n+1} &= y_n + \frac{1}{24}h(55y'_n - 59y'_{n-1} + 37y'_{n-2} - 9y'_{n-3}) \\ z_{n+1} &= z_n + \frac{1}{24}h(55z'_n - 59z'_{n-1} + 37z'_{n-2} - 9z'_{n-3})\end{aligned} \quad (3.38)$$

and

$$\begin{aligned}x_{n+1} &= x_n + \frac{1}{24}h(9x'_{n+1} + 19x'_n - 5x'_{n-1} + 5x'_{n-2}) \\ y_{n+1} &= y_n + \frac{1}{24}h(9y'_{n+1} + 19y'_n - 5y'_{n-1} + 5y'_{n-2}) \\ z_{n+1} &= z_n + \frac{1}{24}h(9z'_{n+1} + 19z'_n - 5z'_{n-1} + 5z'_{n-2})\end{aligned} \quad (3.39)$$

Once $x_{n-3}, x_{n-2}, x_{n-1}, x_n$, etc. are known, one can compute $x'_{n-3}, x'_{n-2}, x'_{n-1}, x'_n$ from (3.36) and then use the predictor formulas (3.38) to obtain a first value of $x_{n+1}, y_{n+1}, z_{n+1}$. Then the predictor formula (3.39) can be used to obtain improved values of $x_{n+1}, y_{n+1}, z_{n+1}$. Of course it is possible to continue to use the corrector formulas (3.39) if the change in y_{n+1} is too large. As a general rule, however, if it is necessary to use the corrector formulas more than once or twice, it can be expected that the step size h is too large and should be made smaller. This formula has a local discretization error no greater than h^5.

A natural question to ask, is why use the predictor-corrector method with a local discretization error proportional to h^5 when the Runge-Kutta method has an error of the same order of magnitude? The answer is that the most consuming part of the calculation is the evaluation of the function $f(x, y, z)$ at successive steps. For the Runge-Kutta method, four evaluations of f are required in taking one step. On the other hand, for the Adams-Moulton method (3.38) and (3.39), the method requires only the evaluation of $f(x, y, z)$ at x_{n+1} for each iteration performed. Thus the number of function evaluations can be reduced significantly.

3.7 The Attitude Control Simulator

The adaptive control architectures described in this book are evaluated by computer simulations. The testbed used, the attitude control problem, requires the simulation of the dynamics of a rigid body. Thus, constructing an accurate simulation of the dynamics of the system, described by equations (3.24) and (3.32), is the first step in the present work.

Using the simulation it is then a simple matter to introduce further complications for the adaptive control system, such as varying the geometry of the system and creating malfunctions of the thrusters. The simulator uses the Adams-Moulton method (section 3.6.2) for the numerical integration of the equations describing the plant dynamics, since these equations are nonlinear systems of ordinary differential equations and no analytical method of solution is available. An analytical solution of the Euler equations for two special cases can be found in Appendix 9.

In the simulator implementation the Adams-Moulton method was preferred over the Runge-Kutta methods (for simplicity reasons in Chapter 5 the Runge-Kutta method is used), because it is computationally less expensive (as it was already noted above). The ANSI C code for an attitude control simulator can be found in Appendix B.

Chapter 4

Artificial Neural Networks

Can Machines Think?
A. Turing
(*Computing Machinery and Intelligence*), Mind vol. LIX

Interest in artificial neural networks (ANN) has grown rapidly over the past few years. This followed a long period of low activity in the field, since Minsky and Papert [142] published their book *Perceptrons* with proofs showing the limitations of the one layer networks. Today, professionals from many diverse fields such as engineering, biology, philosophy and psychology are intrigued by the possibilities offered by this technology, and are seeking applications within their disciplines.

Artificial neural networks are biologically inspired. They are composed of elements that perform in a manner that is analogous to the most elementary functions of the biological neuron. Artificial neural networks exhibit a number of brain characteristics. For example they learn from experience, generalize from previous examples to new ones, and extract essential characteristics from inputs containing irrelevant or noisy data.

Artificial neural networks can modify their behavior in response to the environment. Given a set of inputs (perhaps with desired outputs), neural networks self-adjust to produce consistent responses.

Once trained, a network response can be, to a degree, insensitive to minor variations in its input. This ability to "see" through noise and distortion, and extract the underlying pattern is vital to pattern recognition in a real-world environment. In this way, a model that can deal with an imperfect environment is constructed.

Some artificial neural networks are capable of abstracting the essence of a set of inputs. In a sense, neural networks can learn to produce something they have never "seen" before [119], [207].

The wide applicability of artificial neural networks is one of their most powerful characteristics. They can address many real life tasks. In addition, their possible hardware implementation due to their inherently parallel nature, makes them ideal for real time applications.

4.1 Multilayer Feedforward Neural Networks

The term neural networks usually refers to a linear interconnection of memory-free scalar nonlinear units, supplemented by memory elements (integrators or delay lines) when dynamical behavior is of interest. The coefficients characterizing the connections, called "weights", play a role vaguely analogous to the concentrations of neurotransmitters in biological synapses, while the nonlinear elements in this over-simplified analogy correspond to the neurons themselves.

Layered feedforward networks were called *perceptrons* when they were first studied by Rosenblatt in 1962. In these networks there is a set of input nodes (input layer), whose only role is to feed input patterns into the rest of the network (Figure 4.1). Following the input layer, and before the output layer, there are one or more intermediate layers of units. These units are called *hidden units*, because they have no direct connection to the outside environment, neither input nor output. In the feedforward networks there are not any connections leading from a unit to units in previous layers, nor to other units in the same layer, nor to units more than one layer ahead. The notation used throughout this book will be as in the following example: A network with 10 units in the input layer, 20 units in the hidden layer and 5 units in the output layer will be noted as a 10-20-5 network.

The output of every unit is connected only to the units in the next layer. Every unit is associated with a nonlinear function called the *activation function*. Usually the standard sigmoidal function

$$f(net) = \frac{1}{1 + e^{-net}} \qquad (4.1)$$

is used as the activation function, but other types of functions have been used as

4.1. MULTILAYER FEEDFORWARD NEURAL NETWORKS

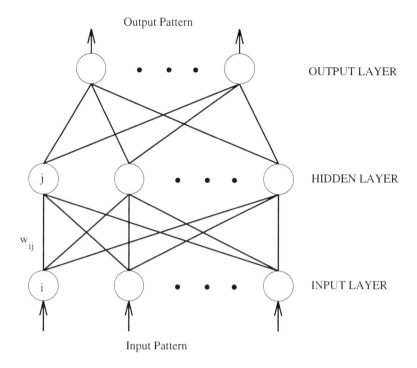

Figure 4.1: An example of a multilayer feedforward neural network.

well [108]. Thus the output of each node j in the network is given by equation (4.1), where its input net is given by the following equation

$$net_j = \sum_i w_{ij} o_i \qquad (4.2)$$

where i runs for all the nodes in the previous layer. The values o_i in the equation are the outputs of the units i in the previous layer, and w_{ij} are the values (weights) associated with the node connections (see Figure 4.1). Associated with supervised learning algorithms is a training set, containing inputs and target outputs for each training input respectively.

Standard feedforward neural networks, with as few as one hidden layer, using (fixed) arbitrary squashing (sigmoidal) functions, can approximate to any desired degree of accuracy, any continuous function $f : \mathcal{R}^n \to \mathcal{R}^m$ over a compact subset of $\mathcal{R}^n$, provided sufficiently many hidden units are available [16], [45], [101]. This is an existence theorem and gives no guarantee that any particular training method (adaptation of the learnable parameters w_{ij}) will converge to the required approximation, nor any indication of the number of hidden nodes required. However, it is an important and significant theoretical result for the application of such networks in various problems[1]. Although one hidden layer is sufficient in most cases, in practice more hidden layers can affect the convergence of a specific learning algorithm (as they change the shape of the multidimensional error surface).

4.2 The Backpropagation Algorithm

Artificial networks learn to perform various tasks by adaptation of their learnable parameters (weights). For this reason a learning algorithm must be used. One of the most widely used is the backpropagation algorithm. Equivalent forms of the backpropagation algorithm were developed by Paul Werbos in 1974 (in his PhD thesis)[216], by David Parker in 1984, and by the PDP group[2] [135], [175], [176].

Feedforward networks have a task associated with the optimization of a defined error function. Usually this is defined as

$$E = \frac{1}{2} \sum_p \sum_k (t_{pk} - o_{pk})^2 \quad (4.3)$$

where k, p belongs to the sets of output nodes and training patterns respectively. The values o_{pk} and t_{pk} are the actual network outputs and the training set target outputs respectively, of the nodes k in the output layer for the training pattern p.

The backpropagation training algorithm is known as the *generalized delta rule* [160], [175] and is based on *steepest descent* optimization methods (see [53] and [205]). The architecture of the network in Figure 4.1 with its weights, and the optimization task (4.3) define an error surface. At each point of the surface, it is desirable to modify the weight vector **w**, so as to decrease the value of the error function (4.3). This can be achieved by gradient descent, following the equation

$$\Delta w_{ij} = -\eta \frac{\partial E}{\partial w_{ij}} \quad (4.4)$$

where $\eta > 0$ is the learning rate of the algorithm, specifying the "speed" of descending the error surface towards a local (global) minimum.

---

[1] Some results describing an upper bound on the required number of nodes in hidden layers, for some special cases, are found in [30].
[2] Parallel Distributed Processing research group at University of California, San Diego.

4.2. THE BACKPROPAGATION ALGORITHM

Using the chain rule of differentiation in (4.4) and substituting from (4.3), (4.1), and (4.2) the following equation for updating the connections (weights) of a network, according to gradient descent backpropagation, is derived:

$$\Delta w_{ij} = \eta \delta_{pj} o_{pi} \tag{4.5}$$

where

$$\delta_{pj} o_{pi} = -\frac{\partial E}{\partial w_{ij}} \tag{4.6}$$

and the subscript p denotes the pattern number. If the j nodes are output-layer nodes

$$\delta_{pj} = (t_{pj} - o_{pj}) f'_j(net_{pj}) \tag{4.7}$$

where t_{pj} denotes the target output according to the training set, for training pattern p, output node j. If j belongs to a hidden unit then

$$\delta_{pj} = f'_j(net_{pj}) \sum_k \delta_{pk} w_{kj} \tag{4.8}$$

where k sums over all the nodes in the next layer and if the function (4.1) is used as the node activation function

$$f'_j(net_{pj}) = \frac{\partial o_j}{\partial net_j} = o_j(1 - o_j) \tag{4.9}$$

Although gradient descent methods suffer from the problem of local minima, in practice when using backpropagation one almost always finds a "good" local minimum. In addition the numerous variations of backpropagation algorithms help in escaping from the local minima problem [210]. The introduction of a *momentum* term often helps in accelerating the training procedure. According to this variation, the weight updates, at time step n, are calculated using the following formula

$$\Delta w_{ij}(n+1) = -\eta \frac{\partial E}{\partial w_{ij}} + \alpha \Delta w_{ij}(n) \tag{4.10}$$

where α is the *momentum* term. This provides a kind of momentum that effectively filters out high-frequency variations of the error surface in the weight space [175].

The success or not of the backpropagation algorithm depends on the generalization capability of the network, i.e the ability to respond correctly for patterns which have not been used in training. Much depends on the architecture (number of nodes and layers) of the network. In general the number of network connections should be kept small compared with the number of training patterns for good generalization results. Some theoretical results on this problem can be found in [90]. A rule of thumb, employed by some researchers, is to use 1/10 of the number of training patterns as the number of network weights.

Another important factor is the time at which the training procedure is stopped. Usually one does not use all of the available problem data as the training data. Instead the available data are split into three sets: the training set, the validation set and the test set. The training set is used for training data, and while training in discrete time periods, the validation set is used to test the current state of the network (the validation set is distinct from the training set). The training error is decreasing continuously, but the error on the validation set will start increasing at some point (Figure 4.2). This is the point that the training procedure should stop, since this indicates the best generalization. Now the test set (distinct from both the training and the validation sets) can be used, to test the capabilities of the network.

4.3 Other Neural Network Paradigms

The backpropagation type network described in the previous section is the most popular neural network architecture nowadays. However, there are several other networks which are very useful in a variety of tasks and applications. A full description of these is outside the scope of this book, but an attempt to cover the operation of the most popular of them is presented.

4.3.1 Hopfield Networks

The Hopfield networks belong to the class of recurrent neural networks, i.e. nets for which feedback connections to nodes of previous layers exist. A four neuron example can be seen in Figure 4.3. Neurons are connected to each other through a weighted link T_{ij}. There are no loop connections $(i \neq j)$, or otherwise $T_{ii} = 0$.

The operation of recurrent networks lies in the evolution of the network state to a stable condition. According to this scheme, given an input pattern, the network will modify its state and after some time it will "relax" in a stable

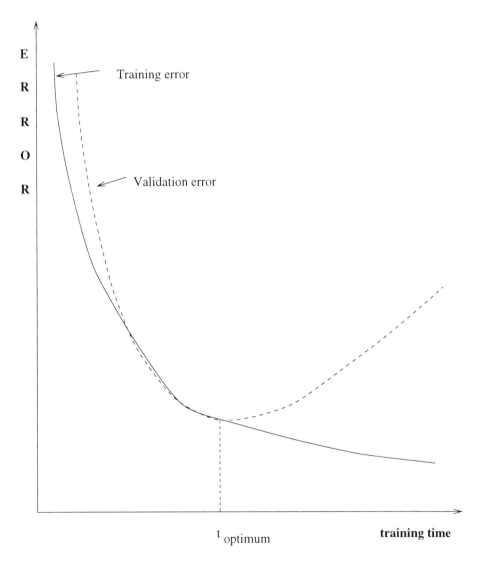

Figure 4.2: An example of when training should be stopped, for good generalization capability of a network.

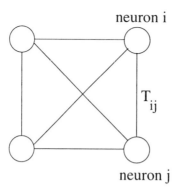

Figure 4.3: An example of a recurrent network. The four nodes are both input and output nodes.

state. Therefore, the key point in these architectures is the ability to construct a network which guarantees some desired stable states. In 1982, John Hopfield [99] introduced an algorithm for the design and analysis of a recurrent network with some embedded stable states. If the system is started in any of these, or similar states, it will end up in one of the stable states.

Such a network can be used as *associative memory* (or content-addressable memory). Incomplete or partially incorrect patterns can be entered in an associative memory network, and then the network will be able to retrieve the full (or even corrected) patterns previously stored in it (the patterns which resemble more the input patterns). Hopfield managed to develop a technique for the design of such associative memory networks. The stability of such networks can be designed by introducing an "energy" function E associated with the network. The stable points correspond to the minima of the energy function. The stability of the Hopfield networks can be proved if it can be shown that the next state of the system will be one of a lower energy than the current state. Thus, the energy function E serves as a Lyapunov function for the system.

The output of neuron i at time t is defined as:

$$V_i(t+1) = \begin{cases} 1, & \text{if } \sum_j T_{ij}V_j(t) > U_i \\ 0, & \text{if } \sum_j T_{ij}V_j(t) < U_i \\ V_i(t), & \text{otherwise} \end{cases} \quad (4.11)$$

where U_i is the threshold associated with neuron i. It should be noted that the Hopfield network operates in an asynchronous fashion.

4.3. OTHER NEURAL NETWORK PARADIGMS

Now define E to be:

$$E = -\frac{1}{2} \sum_{i \neq j} \sum T_{ij} V_i V_j \qquad (4.12)$$

Equation (4.11) causes ΔE (due to ΔV_i) to be always negative:

$$\Delta E = -\frac{1}{2} \Delta V_i \sum_{j \neq i} T_{ij} V_j \qquad (4.13)$$

therefore E is a monotonically decreasing function. The state of the network (defined by the states of all neurons), will continue to change until it reaches a minimum (local or global) for E. This is true only if $T_{ij} = T_{ji}$. If the weights are not symmetric there is no guarantee that the system will settle in a stable state. This case is isomorphic with an Ising model [99].

To store n states V^s, $s = 1 \ldots n$, the following Hebbian rule [85] can be used to assign values to the network weights T_{ij}:

$$T_{ij} = \sum_s (2V_i^s - 1)(2V_j^s - 1) \qquad (4.14)$$

It can be shown that by using (4.14) the stored state would always be stable, if the network states are updated according to (4.11) [99]. However, such a network has a limited capacity in storing "patterns", something which can lead to undesirable effects. Experiments showed that about $0.15N$ patterns can be stored in a network of N neurons, before errors in the pattern retrieval become severe. Some researchers have modified the basic model so as to increase its capacity [52].

If the above model is modified by letting the output V_i of neuron i be either 1 or -1 (instead of 1 and 0), the network states will be symmetric, i.e. if a pattern is stable then its reverse will be stable as well.

The problem with the limited capacity of Hopfield nets leads further to two other problems:

- the stored states can become unstable (if we try to store too many of them).

- spurious stable states (*spin glass states* [84]) can appear (states which do not belong in the set of desired stored patterns) [90].

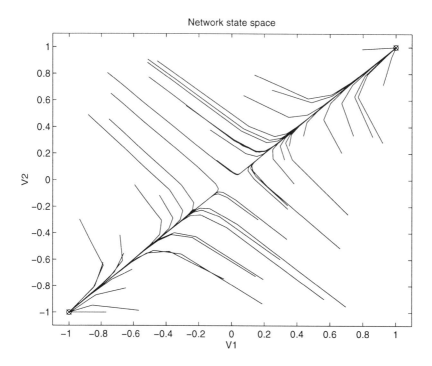

Figure 4.4: An example of the Hopfield network. It is designed so as to store the states $(V_1, V_2) = (1, 1)$ and $(V_1, V_2) = (-1, -1)$. Circles o, denote the stored stable patterns and the cross x denotes the settle point for various initial conditions. All of them, end up in one of the two desired stable states.

The existence of spurious stable states corresponds to local minima of the energy function E.

Figure (4.4) shows an example of a Hopfield network with two neurons. It is designed so as to store the $(V_1, V_2) = (1, 1)$ and $(V_1, V_2) = (-1, -1)$ states. From whatever initial condition the network is started, it will settle in one of the two stored patterns. If an incorrect pattern (initial condition) is given to the network, the correct stored pattern will be retrieved (most of the time but not always [86]). Figure (4.5) displays the energy map, with the corresponding vector field, for the same system. From this, it is obvious that the only stable states are the stored ones. It is worth noting that if the system is started from the initial condition $(V_1, V_2) = (0, 0)$, it will relax in the $(V_1, V_2) = (1, 1)$ state.

However, the attempt to store the states $(V_1, V_2) = (1, 1)$, $(V_1, V_2) = (1, -1)$ and $(V_1, V_2) = (-1, 1)$ leads to the situation of Figure (4.5), where some initial conditions (the ones shown) lead to an undesirable stable point $(V_1, V_2) =$

4.3. OTHER NEURAL NETWORK PARADIGMS

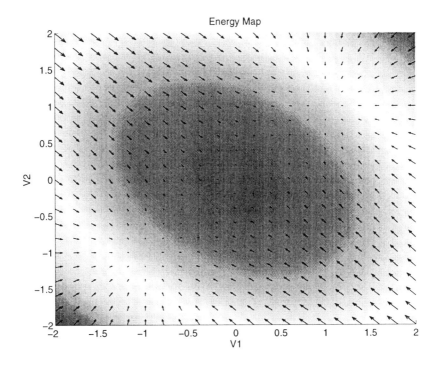

Figure 4.5: The energy map and the vector field for the Hopfield network designed so as to store the states $(V_1, V_2) = (1, 1)$ and $(V_1, V_2) = (-1, -1)$. All initial states lead to the stored stable states as indicated by the arrows of the vector field.

$(-1, -1)$. The spurious stable states are not necessarily symmetric to the stored ones, as was the case with this particular example. In general, spurious stable states can exist which have no correlation with the stored patterns. Some research has concentrated on an attempt to minimize the number of spurious stable states.

The discrete Hopfield network (where neurons can be in one of two states, either *on* or *off*) can be extended to a continuous Hopfield model [100]. Such a model is more similar to the real nervous system (it is generally accepted that real neurons are not binary [8]), and the output (activation) of each neuron is the weighted sum of the excitatory and inhibitory signals to it, transformed by a sigmoidal nonlinearity similar to that of (4.1). The same logical sequence is applied for the analysis of the continuous version: an energy (Lyapunov) function is shown to decrease as the system evolves over time [100].

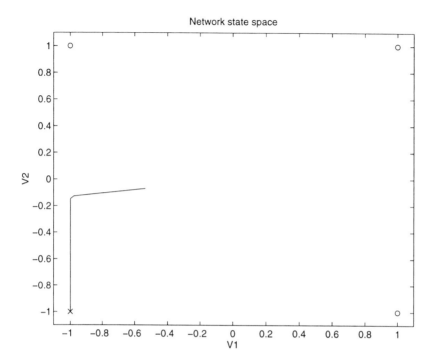

Figure 4.6: Spurious stable states in the Hopfield network. Circles o, denote the stored stable patterns and the cross x denotes the undesired stable state. Given the initial network state shown, the network will settle in the "spurious" stable state.

4.3.2 Boltzmann Machines and Simulated Annealing

As it was discussed in the previous section, Hopfield nets often converge to local (instead of global) minima and this fact gives motivation for the development of new architectures. *Boltzmann machines* [2], [94], [95], are an extension to the discrete model of Hopfield networks to include hidden neurons (Figure 4.7). The weights of the network are also symmetric $T_{ij} = T_{ji}$, but the neurons are updated in a stochastic rather than a deterministic way. In general, Boltzmann machines can be applied for learning to any stochastic network.

The probability of the system being in a state S_i is given by the Boltzmann distribution of statistical mechanics (hence the name Boltzmann machines). The operation of such a network is based on the *simulated annealing* algorithm [1]. This algorithm is derived from the physics principles of metallurgical annealing where a metal is heated to near its melting point and then cooled down. It was first introduced in [138] in the context of Monte Carlo methods

4.3. OTHER NEURAL NETWORK PARADIGMS

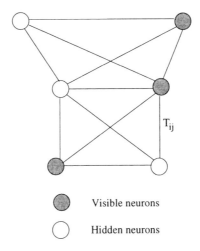

Figure 4.7: The Boltzmann machines consist of visible and hidden neurons interconnected.

and has also been used in other applications, such as "classical" optimization of a function $f(x)$ [117]. The pseudocode in Figure 4.8 describes the simulation annealing procedure, applied to the optimization of a function $f(x)$.

Boltzmann machines update the state of their nodes probabilistically. A node i fires (i.e. takes the value of $+1$) with probability p_i given by (4.15).

$$p_i = \frac{1}{1 + e^{-2n_i/T}} \tag{4.15}$$

where

$$n_i = \frac{1}{2} \sum_{j \neq i} T_{ij} V_j - U_i \tag{4.16}$$

It can be shown that the term n_i corresponds to half of the energy change ΔE_i when the neuron i modifies its output to fire ($V_i = 1$) from a previous non-firing state ($V_i = -1$). Indeed, if the energy is defined as

$$E = -\frac{1}{2} \sum_{\substack{i \\ i \neq j}} \sum_j V_i (T_{ij} V_j - U_i) \tag{4.17}$$

then

$$\Delta E_i = -\frac{1}{2} V_i \sum_j (T_{ij} V_j - U_i) - \frac{1}{2} V_i \sum_j (T_{ij} V_j - U_i) = -2 \cdot n_i \tag{4.18}$$

since the previous state of neuron i was $V_i = -1$.

```
x = random();
T = high;
α ∈ [0.5, 1.0];
initialize M, N : (M < N);
while f(x) has been decreased more than ε in the last L iterations do
        x' = modify(x);
        Δf = f(x') - f(x);
        total_changes = total_changes + 1;
        if Δf < 0 then
                x = x';
                decreasing_changes = decreasing_changes + 1;
        else
                x = x' (with probability e^(-Δf/T));
                if decreasing_changes ≤ M or total_changes ≤ N
                T = αT;
                decreasing_changes = 0;
                total_changes = 0;
                endif
        endif
endwhile
```

Figure 4.8: Pseudocode of the simulation annealing algorithm for the optimization of a function $f(x)$.

The operation network is initially started at a high temperature, which is reduced every time a *thermal equilibrium* is reached. A state of thermal equilibrium for temperature T is one for which the probabilities $p(S_i)$ of the network being in a particular state S_i remain approximately constant (i.e. if the network is allowed to evolve more in time, all $p(S_i)$ will not be altered significantly).

Boltzmann machines belong to the class of supervised learning algorithms. The training set consists of n states $S_1, S_2 \ldots, S_n$ for the visible nodes and n probabilities $P(S_1), P(S_2), \ldots, P(S_n)$ which specify the probability of the network being in one of the n states, when it runs freely (there is no clamping of values in any of the nodes).

To train the network it is desirable to minimize the distance between the training set probabilities $P^+(S_1), P^+(S_2), \ldots, P^+(S_n)$, and the actual proba-

4.3. OTHER NEURAL NETWORK PARADIGMS

bilities $P^-(S_1)$, $P^-(S_2)$, ..., $P^-(S_n)$ of the network being in states S_1, S_2, ..., S_n when it is running free. The error function used in training is given by

$$G = \sum_{\alpha=1...n} P^+(S_\alpha) \ln \left[\frac{P^+(S_\alpha)}{P^-(S_\alpha)} \right] \qquad (4.19)$$

To minimize G, the gradient descent technique is applied to modify the weights of the network:

$$\Delta T_{ij} = -\eta \frac{\partial G}{\partial T_{ij}} \qquad (4.20)$$

it can be shown that

$$\frac{\partial G}{\partial T_{ij}} = -\frac{1}{T}(p_{ij}^+ - p_{ij}^-) \qquad (4.21)$$

where p_{ij}^+ is the average probability of neurons i, j firing simultaneously, when the visible units are clamped to the environment (training set), and p_{ij}^- is the average probability of neurons i, j firing simultaneously when the network is running free. From (4.20) and (4.21) the following rule is obtained for modification of the weights T_{ij} during training:

$$\Delta T_{ij} = \frac{1}{T} \cdot \eta (p_{ij}^+ - p_{ij}^-) \qquad (4.22)$$

Markov chains can facilitate the analysis of operation of Boltzmann machines.

Unlike the Hopfield network which usually converges to a local minimum, Boltzmann machines stop their operation at the global minimum of E (or very close to the global minimum). The association of probabilities with system states introduces "noise" in the network, enabling it to escape from local minima [4]. However, the whole operation of the network could be quite slow depending on the complexity of the error surface, something which is the main drawback for its application. However, the Boltzmann machine has been applied to a variety of problems, including combinatorial optimization problems (the traveling salesman problem (TSP), the graph coloring problem and others) [1].

4.3.3 Unsupervised Competitive Neural Networks

In contrast with the neural network architectures described in the previous sections, there are also networks which follow the unsupervised learning paradigm.

According to this, the training patterns are associated with inputs only, and not outputs (the desired outputs are not known, as is the case with supervised learning). Thus, the training set does not consist of pairs *(inputs, desired outputs)*. The networks which perform unsupervised learning are able to perform

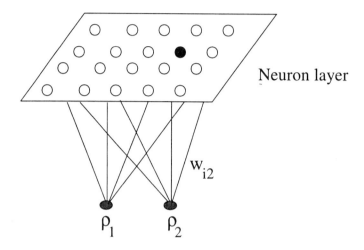

Figure 4.9: The input-patterns to output-nodes mapping created by the Kohonen network.

classification or extract features from the input patterns provided to them. This is based in a measure of similarity of the input patterns. Such networks are also called *feature maps*.

Kohonen networks [119] create a topological mapping of an N dimensional continuous valued input space to an M dimensional continuous valued output space. For example in Figure 4.9, a two-dimensional region described by pairs of values (ρ_1, ρ_2) is mapped onto a two-dimensional space (neuron space). The m output nodes (neurons) are fully connected with the n inputs via weights w_{ij}, $i = 1, \ldots, m$, $j = 1, \ldots, n$.

During training, each time an input pattern ρ, $(\rho = (\rho_1, \rho_2)$ in the example of Figure 4.9) is presented to the network, a single neuron i^* is chosen, according to the distance of its weights w_{ij} from the input ρ:

$$|w_{i^*} - \rho| \leq |w_i - \rho|, \quad \forall i \qquad (4.23)$$

This neuron is said to have won the competition with the other neurons, and together with the neurons in its neighbourhood, it is able to update its weights.

4.3. OTHER NEURAL NETWORK PARADIGMS

The weight updates are made according to Kohonen's rule [119]:

$$\Delta w_{ij} = \eta \cdot \Omega(i, i^*, t)(\rho_j - w_{ij}) \qquad (4.24)$$

where $\Omega(i, i^*, t)$ is a function describing the size of neighborhood of neuron i^* for which neuron weights will be updated, according to their distance from i^*. Thus a neuron i which is far away from i^* will modify its weight only by a small amount. For $i = i^*$, $\Omega = 1$. $\Omega(i, i^*, t)$ is also a function of time t, and is decreasing over time, i.e. as the network progresses, weight updates will be made only for a small neighborhood of the winning node i^*, and eventually only for i^*.

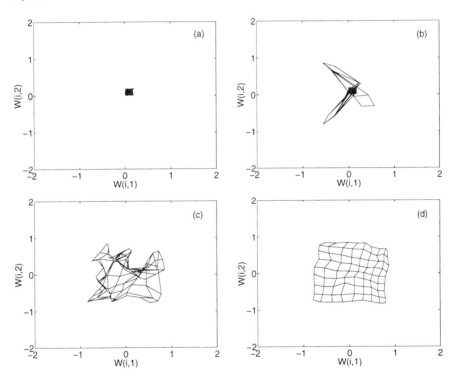

Figure 4.10: A Kohonen network evolves, when it is provided with a 1000 2-input patterns uniformly distributed in the range $[-1, +1]$. (a) Initial state of the network. The initial weights are small so they occupy only a small region of the whole space. (b) Network after 10 iterations. (c) Network after 200 iterations. (d) Network after 3000 iterations.

The above procedure of training the network aims to bring the network weights as close as possible to the input patterns ρ_j. Therefore the network can be thought of as an elastic net [90] which modifies its shape in order to resemble that of the space defined by the input patterns. For example, Figure 4.10 shows the evolution of a Kohonen network when it is provided with a 1000 2-input patterns $\rho = (\rho_1, \rho_2)$ uniformly distributed in the range $[-1, +1]$, i.e. $-1 \leq \rho_1 \leq +1$, $-1 \leq \rho_1 \leq +1$. After 3000 iterations, the network resembles the "shape" of the input space. Allowing for more iterations, this resemblance will become even greater.

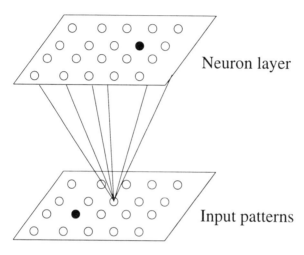

Figure 4.11: Willshaw and von Malsburg architecture for competitive learning.

A similar type of network is that of *competitive learning networks* (although broadly speaking one could say that Kohonen networks belong to this family). The principles here remain the same, but the difference with the Kohonen networks is that only the neuron i^* winning the competition against the other neurons is allowed to modify its weights $w_{i^* j}$. One of the earliest works in this type of network was that of Willshaw and von der Malsburg [226] who used the model shown in Figure 4.11. This architecture is of biological interest for describing the way that various mappings are performed in biological organisms (e.g. retina to cortex mapping).

4.3.4 Adaptive Resonance Theory (ART) Networks

Competitive learning networks suffer from the well known *stability-plasticity dilemma* of learning systems [35]. Unless the learning rate η is gradually decreased to zero (allowing no more changes in the network, i.e. no network plasticity exists any more), the decisions of the network regarding the category

4.3. OTHER NEURAL NETWORK PARADIGMS

to which a pattern belongs may keep changing.

Learning a new "environment" (plasticity) without forgetting previous "experiences" is a fundamental desirable property for real learning systems. A new architecture, the *adaptive resonance theory (ART)*, was developed for this purpose [33], [34]. The networks based on this theory categorize input patterns only if they are sufficiently similar (*resonate*) with a stored exemplar $\mathbf{w_i}$ of a category. In this case, the stored exemplars adapt so as to provide "good" recognition codes for the corresponding category. If an input pattern does not match an exemplar sufficiently, a new category (and exemplar) is created by the use of a previously uncommitted output node. If no uncommitted node is available any more, the ART network gives no response, thus preserving its stability. The sufficiency of similarity between two patterns depends on a *vigilance parameter* ρ, where $0 < \rho \leq 1$.

Adaptive resonance theory networks are based on the algorithm shown in Figure 4.12. Initially all output nodes i are uncommitted (they define no cat-

```
w_i := 1,  ∀i;
enable all nodes i ;
found := 0;
w̄_i := w_i/(ε + ∑_j w_ij),  ∀i ;
while found = 0
        i* = i :   max(w̄_i · I), i ∈ enabled_nodes ;
        r := w_i* · I/(∑_j I_j) ;
        if r ≥ ρ
            w_i* := w_i* AND I ;
            found := 1 ;
        else
            reject exemplar w_i* ;
            disable node i* ;
        endif
endwhile
```

Figure 4.12: Pseudocode of the learning algorithm on which ART networks are based.

egory). This is denoted as $\mathbf{w_i} = \mathbf{1}$. Every time a new pattern $\mathbf{I}$ is presented to the network, all output nodes Y_i are enabled (Figure 4.13). A winner i^* is chosen based on the fraction of bits in exemplar $\mathbf{w_i}$ that are also in the input pattern **I**. This is measured by $\bar{\mathbf{w}}_\mathbf{i} \cdot \mathbf{I}$, where $\bar{\mathbf{w}}_\mathbf{i}$ is a normalized form of $\mathbf{w}_\mathbf{i}$ defined by

$$\bar{\mathbf{w}}_\mathbf{i} = \frac{\mathbf{w}_\mathbf{i}}{\epsilon + \sum_j w_{ij}} \tag{4.25}$$

where ϵ is chosen so that if two exemplars have the same $\bar{\mathbf{w}}_\mathbf{i} \cdot \mathbf{I}$, the longer one will be chosen as the winner i^*. Following this, the fraction of bits r in **I** which are also in $\mathbf{w}_{\mathbf{i}^*}$ is calculated by

$$r = \frac{\mathbf{w}_{\mathbf{i}^*} \cdot \mathbf{I}}{\sum_j I_j} \tag{4.26}$$

If **I** is sufficiently similar with $\mathbf{w}_{\mathbf{i}^*}$ (there is resonance), i.e. $r \geq \rho$ the category defined by exemplar $\mathbf{w}_{\mathbf{i}^*}$ is accepted and $\mathbf{w}_{\mathbf{i}^*}$ is further adjusted, by deleting (masking) any bits which are not also in **I**. If $r < \rho$, the input pattern **I** does not belong to the category defined by exemplar $\mathbf{w}_{\mathbf{i}^*}$. Node i^* is disabled, and a next candidate exemplar $\mathbf{w}_{\mathbf{i}'^*}$ is chosen based on the next largest value of $\mathbf{w}_{\mathbf{i}^*} \cdot \mathbf{I}$ among all remaining enabled nodes i. The same steps as before are repeated, until a category i is found for input **I**, or until there are no more enabled nodes, in which case there is no network decision.

Note that the creation of new categories is possible, when the winning node i^* is uncommitted, i.e. $\mathbf{w}_{\mathbf{i}^*} = \mathbf{1}$, in which case it modifies itself by masking all bits which are not also in **I**. The vigilance parameter ρ defines the "resolution" of the categories which will be created. Larger values of ρ increase both the resolution and the number of categories formed. Versions of ART networks which use adaptive ρ (changing during learning) are possible [90].

The implementation of the above algorithm in a network is shown in Figure 4.13. The network consists of two main competitive layers: the comparison layer with nodes V_j, and the output (recognition) layer with nodes Y_i. The two layers are fully connected in both directions, $\bar{w}_{ij}$ (as defined in (4.25)) being the weights of the connections from the comparison layer to the output layer, and w_{ij} being the weights of the backwards connections (w_{ij} are binary). An input **I** arrives in the comparison layer. The comparison layer sends a signal vector **S** to the output layer, so that the node i^* with the largest input $\sum_j \bar{w}_{ij} V_j$ among all enabled nodes will be activated, i.e. $Y_{i^*} = 1$. Following this, the output layer generates a signal vector **U** which is sent down to the comparison layer, changing the activation of nodes V_j according to the following equation:

$$V_j = \begin{cases} I_j, & \text{if } Y_i = 0, \forall i \\ I_j \wedge \sum_i w_{ij} Y_i, & \text{otherwise} \end{cases} \tag{4.27}$$

4.3. OTHER NEURAL NETWORK PARADIGMS

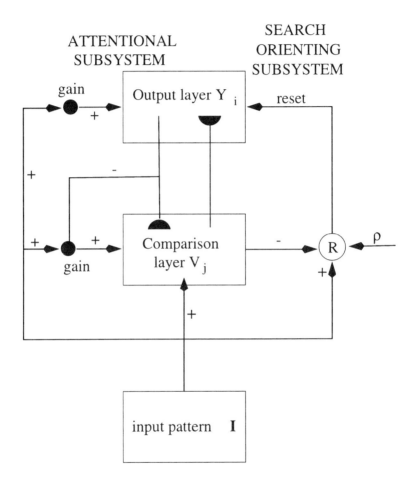

Figure 4.13: A typical ART1 architecture.

where $\land$ is the logical *AND* operation. Thus the template pattern **V** is associated with the current active category [36]. If the similarity of patterns **I** and **V** is not sufficient, a *reset* signal "shuts off" the current active category and the whole sequence of operations is started again. If the current active category is accepted, its weights w_{ij} are updated according to:

$$\frac{dw_{ij}}{dt} = \eta \cdot Y_i(V_j - w_{ij}) \qquad (4.28)$$

so the exemplar weight w_{i^*j} for the winning node i^* is adaptively modified by "masking out" the bits which are not also in **I**. Similarly to the other learning algorithms previously described, η represents the learning rate for the adaptation of w_{ij}.

Because patterns of activity in the two layers can persist only for a short time, the pattern representations in these layers can be seen as *short term memory (STM)*. On the other hand, the connection weights, which can be modified by inputs and persist for very long times after input offset can be seen as *long term memory (LTM)* [36].

The adaptive resonance theory networks comprise a whole family of networks (all of them based on similar basic principles) and the above description with Figure 4.13 is an outline of the operation for the ART1 model introduced in [34]. An extension of ART1 to work with continuous value inputs is the ART2 [33]. Like ART1, the ART2 architecture consists of two main layers (the comparison and the recognition layer), but now the comparison layer can be further decomposed in three processing layers. The problem of asymmetry in the design of the comparison and output layers is solved in the ART3 model, where a new search algorithm is incorporated [36]. Other members of the ART family networks include the Fuzzy ART, ARTMAP, Fuzzy ARTMAP and Fusion ARTMAP [37]. A recent development in the family is that of dART (Distributed ART) which allows an arbitrarily distributed code representation [32].

4.3.5 Radial Basis Function Networks

The *radial basis function* networks (introduced in [28]), are feedforward networks with three main differences from the backpropagation-type feedforward networks (multilayer perceptrons (MLP)):

- they always consist of only one hidden layer.
- the output nodes are always linear.

4.3. OTHER NEURAL NETWORK PARADIGMS

- the hidden nodes have radial basis functions (RBF) $\varphi_i(\mathbf{x})$ as activation functions, rather than sigmoidals.

The radial basis functions $\varphi_i(\mathbf{x})$ usually take the Gaussian form:

$$\varphi_i(x_1, x_2, \ldots, x_n) = e^{-[(x_1-c_{1i})^2+(x_2-c_{2i})^2+\ldots+(x_n-c_{ni})^2]/\sigma_i^2} \qquad (4.29)$$

where $\mathbf{c}_i = (c_{1i}, c_{2i}, \ldots, c_{ni})$ is a vector which defines the center of the radial basis function φ_i in neuron i, and σ_i is the "shape" of the function. Input patterns $\mathbf{x}$ activate the nodes according to their distance $||\mathbf{x} - \mathbf{c}_i||$ from the node centers $\mathbf{c}_i$. Thus, each hidden neuron responds only to inputs which are in a region (*receptive field*) around its center $\mathbf{c}_i$.

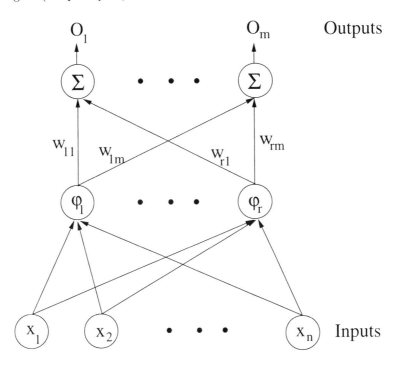

Figure 4.14: A typical RBF network.

Other functions can also be used as the activation functions of the radial basis nodes, without significantly affecting the performance of the RBF networks [84]. The architecture of a RBF network is shown in Figure 4.14. The outputs O_j, $j = 1, \ldots, m$ are the sum of the linear combination of the radial basis function outputs $\varphi_i(\mathbf{x})$, with the weights w_{ij} of the connections from the hidden to the output nodes:

$$O_j(\mathbf{x}) = \sum_{i=1}^{r} w_{ij} \cdot \varphi_i(\mathbf{x}) \qquad j = 1, \ldots, m \qquad (4.30)$$

Training of the RBF networks consists of two parts: an unsupervised part (any vector quantization or competitive learning algorithm) to define of $\mathbf{c}_i$ and σ_i, and a supervised part to "learn" the weights w_{ij}:

$$\Delta w_{ij} = \eta [t_j^{\mathbf{x}} - O_j^{\mathbf{x}}] \varphi_i(\mathbf{x}) \qquad (4.31)$$

where $t_j^{\mathbf{x}}$ and $O_j^{\mathbf{x}}$ are the target output and the actual output of node j respectively, upon presentation of the input pattern $\mathbf{x}$.

It has been shown (see [84], for example) that given a sufficient number of radial basis nodes, any function can be constructed using a RBF net. Radial basis function networks are *local learning networks* since they are able to construct good approximations, only in the regions where a sufficient number of training data is provided. In contrast, multilayer perceptrons are *global learning networks* constructing global approximations of the nonlinear input-output space. Local networks are able to learn much faster when the number of inputs is small. However, when the number of inputs increases, MLP networks give much better generalization [210]. This can easily be seen by noticing that local approximations are in a sense similar to look-up tables. As the number of the function inputs increases, the size of the look-up table (radial basis functions) increases exponentially.

Other more powerful designs following the supervised learning paradigm, include those of time-lagged recurrent networks (TLRN) and the most recent development of simultaneous recurrent networks (SRN) [210], [217], [219].

Chapter 5

Neuromodels of Dynamic Systems

Let the biologists go as far as they can and let us go as far as we can. Some day the two will meet.
Sigmund Freud
(*Origins of Psychoanalysis*)

Nearly all phenomena of the natural world involve systems whose behavior varies through time. In some cases the rules governing the behavior are themselves opaque, but in many cases complexity can arise from relatively simple rules. The most primitive biological information processing systems evolved to meet the necessities of survival. From sense data to action, flight or the capture of prey, there is a gap that was bridged by the evolution of adaptive control systems based on circuits of simple neural components. A vital computational characteristic of such neural circuitry is the ability to model non-linear dynamic systems.

The mathematical description of dynamic systems is well established, using vector differential or difference equations. For example, a continuous-time dynamic system, can be written as:

$$\frac{d\mathbf{x}}{dt} \stackrel{def}{=} \dot{\mathbf{x}}(t) = f(\mathbf{x}(t), \mathbf{c}, t) \,. \tag{5.1}$$

where $\mathbf{x}(t) \in \mathcal{R}^n$, $\mathbf{c} \in \mathcal{R}^m$ and f is a mapping defined as $f : \mathcal{R}^n \times \mathcal{R}^m \to \mathcal{R}^n$. According to the discussion in Chapter 2, the vector $\mathbf{x}(t)$ is referred to as the *state* of the system at time t.

When a dynamic system like (5.1) is modeled, a discrete system

$$\mathbf{x}'(t_{k+1}) = \mathbf{g}(\mathbf{x}'(t_k), \mathbf{x}'(t_{k-1}), \ldots, \mathbf{x}'(t_{k-p}), \mathbf{c}) \tag{5.2}$$

is constructed, such that the values $\mathbf{x}'(t_k)$ approximate the solution $\mathbf{x}(t_k)$ of (5.1) at the time sequence $t_0 < t_1 < t_2 < \cdots t_n < \cdots$. There is a clear relationship between modeling and prediction.

The profound significance of dynamic system modeling (DSM) in areas like signal processing, aerodynamics, nuclear power generation, physiology, and adaptive control (where DSM is called plant identification) makes DSM a very active research area.

Although diverse representations (Wiener series, Volterra series, Uryson series) of some classes of nonlinear dynamic systems have been derived from the Stone-Weierstrass theorem [174] (the generalization of the well-known Weierstrass theorem which states that any continuous function over a compact interval can be uniformly approximated by polynomials), their application to many non-linear problems has often not been successful.

In biological control systems the success or failure of the control action is constantly checked using incoming sense data. In this work, a neural network, which acts in this way, will be called a *locally predictive net* (LPN) (similarly, some control theorists call this a *series-parallel plant identification model* [148]). A series of experiments is used for training networks (by backpropagation through error) which look at the current state $((x(t), \dot{x}(t))$ and at a number of past states $(x(t - \Delta t), \dot{x}(t - \Delta t), x(t - 2\Delta t), \dot{x}(t - 2\Delta t), \ldots, x(t - p\Delta t), \dot{x}(t - p\Delta t))$ of the actual dynamic system, and attempt to predict the next state $(x(t + \Delta t), \dot{x}(t + \Delta t))$ of the system (similar approaches were taken in [125], [28]). Thus we seek to construct a network to implement the mapping

$$(x(t), \dot{x}(t), \ldots, x(t - p\Delta t), \dot{x}(t - p\Delta t)) \longrightarrow (x(t + \Delta t), \dot{x}(t + \Delta t)) \tag{5.3}$$

for some fixed $p \geq 1$. Of course, there is no particular reason why the sampling interval need be the same as the prediction interval Δt.

While empirical tests and simulations show such structures to be effective identification models, it is not clear how universal these models are, and whether they can describe the input-output behavior of general dynamical systems. The flow ϕ_t of a chaotic system, which in principle occurs in a high-dimensional space Λ, evolves to a compact attracting manifold Λ^* of low dimensionality d. The chaotic attractor Λ^* often has a *fractal* structure, for which d is not an integer [134]. It has been shown [157], [198] that for a general class of nonlinear systems, there exists a smooth function ψ that satisfies the mapping (5.3) in a restricted region of operation around an equilibrium point, with $p \in (d-1, 2d)$. Recently in [128], using some fundamental concepts from control theory and differential topology, the existence of global input-output models was established, i.e. for any member of the class, (5.3) holds for some finite p.

In this work, the capability of neural networks in modeling dynamic systems is investigated. By "modeling a dynamic system", the interest is concentrated in relatively accurate prediction of analytically smooth systems, for a short time ahead, in the case where there is continuous sensory update. In contrast with other approaches, data for all the state variables are used for prediction, instead of just one, and a single trajectory of the dynamic system is used for the training of a neural network.

Moody and Darken [144] pointed out, that if an equispaced set of samples Δt is taken, and we try to predict an interval T ahead, then classical techniques like global linear autoregression, or Garbor-Volterra-Wiener polynomials expansions typically fail for $T > t_{char}$, where t_{char} is the inverse of the mean of the power spectrum. They give an example for the Mackey-Glass equation of a localized field network, that is able to achieve a mean-squared error $E \approx 0.05$, for $T = 85 \approx 1.7 t_{char}$. The experience from the experiments in this book suggests that it is easier to obtain reasonably accurate long term prediction by training a network to predict in a single step than by training it for short term prediction, and then iterating it. Such iteration tends to lead to a rapid accumulation of errors.

The following examples suggest that in many cases, training on a single trajectory can lead to a network which has a good local predictive capability (the embedding map) over the whole, or a large region, of the phase space. This implies a surprising generalization capability for the LPN networks. It is not known yet how robust this phenomenon is. The analogy which springs to mind, is analytic continuation: is there an underlying C^∞ approximation theorem here? For example, in the case of the Van der Pol system, a network trained on a trajectory external to the limit cycle (the trajectory therefore spirals inwards), nevertheless learns a good model for an internal trajectory (which spirals outwards). What can be said at present is that the examples in this chapter suggest the existence of such a phenomenon. This may be a reflection of self-similarity properties of the input-output model (5.3) rather than a property associated with the neural networks themselves.

In this chapter, four different dynamic systems of increasing complexity are considered: a two point attractor system, a system with a limit cycle, a chaotic system, and the rotational motion of a rigid body under a chaotic control regime. For each system we construct a LPN neuromodel, and illustrate graphically, the extent to which the neuromodel is an accurate local predictor for the dynamic system. Typically, a few hundred training points from a single trajectory have been used, since this is computationally convenient and serves to demonstrate the generalization capability of the LPN during testing. However, experiments with the same number of training points, uniformly distributed over the phase space, indicated similar results.

Considerable care is needed to calculate the evolution of the dynamic systems accurately. The Runge-Kutta method described in Chapter 3 has been used. Whilst requiring high precision numerical accuracy, the overall aim is to give an essentially qualitative description of the modeling phenomenon, and so usually details of units are ignored. However, time is quantified in seconds. The LPN predictions proceed, for the most part, in steps of Δt, in the range $[0.1, 1]$ sec, and the notation $t = k\Delta t$ ($k = 0, 1, 2, \dots$) where the initial value is $t = 0$.

5.1 A Pattern Recognition Task

One very straightforward way to use neural networks in this context would be as follows. Consider a simple dynamic system with several point attractors, and attempt to train the neural net so that, given an initial state of the system, the network can predict which attractor the trajectory will eventually reach (an example of a two point attractor system is shown in Figure 5.1: depending on the initial height that the ball will be left, after some time it will rest in one of the two local minima).

For example, consider the damped oscillator system

$$m\ddot{x} + c\dot{x} - \alpha x + bx^3 = 0 \tag{5.4}$$

This system has equilibria at $x = 0$ and $x = \pm\sqrt{(a/b)}$. Equation (5.4) models

5.1. A PATTERN RECOGNITION TASK

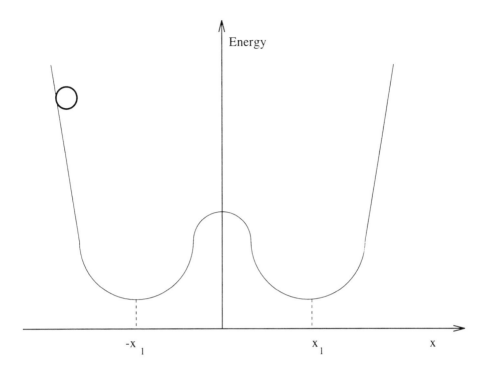

Figure 5.1: A system with two point attractors. The ball after some oscillations will rest in either of the attractors, depending on its initial energy.

the motion of a mass in a potential field exhibiting two minima. If $c = 0$, there is no damping and there are two kinds of trajectories: those that cross the potential barrier repeatedly, and those that simply orbit one of the attractors. If c is small but positive, the system is lightly damped [106]. The points at $x = \pm\sqrt{(a/b)}$ are competing point attractors, and in this symmetric problem half of the phase space $(x, \dot{x})$ leads to one attractor, while the other half leads to the other. There is an unstable equilibrium at $x = 0$. For these experiments the following values were used: $m = 2.5, c = 2.0, a = 3, b = 4$ so that $\pm\sqrt{(a/b)} \approx \pm 0.866$.

Using the substitution $\dot{x} = y$ the differential equation (5.4) can be transformed into a system of first order differential equations:

$$\dot{x} = y \qquad (5.5)$$
$$\dot{y} = -\frac{cy}{m} + \frac{ax}{m} - \frac{b}{m}x^3$$

in which form a numerical algorithm, such as the Runge-Kutta method, can be applied.

Figure 5.2 shows the phase space $(x, \dot{x})$ for this system, with the initial points colored black if the trajectory is attracted to the righthand attractor, and white if the trajectory is attracted to the lefthand attractor. The diagram was constructed by choosing initial points arranged on a grid, and following each trajectory using the Runge-Kutta method. (It is worth observing that for this particular dynamic system, there is a much easier method to determine the boundary between the black and white zones. One can linearize the system about the unstable attractor at $x = 0$. Solving the linear system gives a good approximation of the direction of the trajectory through this point, and iterating this trajectory gives the *separatrix*). In this pattern recognition task, a 2-10-10-1 network was trained using backpropagation, with a learning rate $\eta = 0.2$ and a momentum $\alpha = 0.7$. The values for $(x, \dot{x})$ over the range $x \in [-2, 2]$ and $\dot{x} \in [-1, -1]$ were scaled to $[0, 1]$, and applied to the two inputs. The training data were scaled to avoid arithmetic overflow problems that occur in cases where the weights of a network become too large. In addition the scaling of the training set helps to avoid the problem of *network paralysis*. The network paralysis problem occurs when the weights of the network become too large. In this case the sigmoidal activation functions operate on their flat region, and so the error derivatives are extremely small. Consequently the weights are updated by only a very small value, and thus the network "paralyzes" (remains) at the same point of the error surface.

In this experiment the output node was used to predict the attractor to which the initial condition in the input nodes leads. In this way the net was trained to predict the correct attractor.

The input training data consisted of 861 initial points, arranged as a 41×21 grid (step 0.1), and the corresponding output values $\{0, 1\}$ associated with each attractor. Training continued for 40,000 iterations. When the training was complete, the resulting network was used to get a high resolution image of $80,601 = 401 \times 201$ points (step 0.01), (Figure 5.3).

This is a rather simple way to use a neural network to predict a dynamic system since it is nothing more nor less than direct pattern recognition. As one might expect it works reasonably well using the outcome of 861 distinct trajectories, to produce quite accurate results near the center of the region, but these become somewhat less reliable near the boundaries. However, when compared with the granularity of the training data, it can be seen that the interpolative power of the network is quite high.

5.1. A PATTERN RECOGNITION TASK

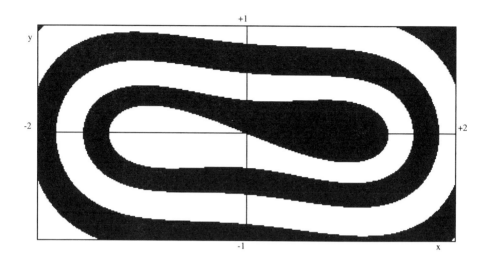

Figure 5.2: The phase space for the 2-attractor system, colored black for points which converge to the right attractor and white for points which converge to the left attractor.

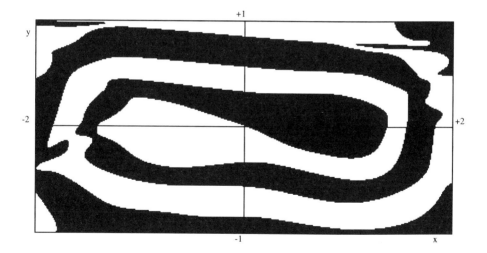

Figure 5.3: A 2-10-10-1 pattern recognizing net attempting to predict the target attractor.

5.2 Local Predictive Networks (LPN) Applied to Dynamic System Modeling

The methodology of direct pattern recognition plainly has very limited utility when applied to more complex dynamic systems. We therefore turn to the issue, posed in the introduction, of getting a network to form an input-output model of the dynamic system.

The choice of Δt depends on the complexity of the dynamic system under consideration. For simpler systems Δt can be large, giving a relatively long term prediction, whilst maintaining accuracy. The choice of Δt must be combined with an appropriate choice of the size of the network, in order to obtain good results. For small Δt, the relative change of system state is very small, and so a larger network is needed for accurate modeling. As Δt increases, the network size required for the same relative accuracy first decreases. For even larger Δt the required mapping increases in complexity and this is reflected in the fact that again a larger network size is required for the LPN.

The complexity of a dynamic system can be described in general terms by its Lyapunov exponents [227], which measure the sensitivity of the system to changes in the initial conditions. The Lyapunov exponents provide a coordinate-independent measure of the asymptotic local stability of properties of a trajectory. Given a continuous dynamic system in an n-dimensional phase space, the evolution of a infinitesimal ball of radius $\epsilon(0)$ centered on a point $\mathbf{V}(0)$ in the phase space is examined, and its progressive deformation as time unfolds. This deformation is governed by the linear part of the flow. The ball thus remains an ellipsoid. Suppose the principal axes of the ellipsoid at time t are of length $\epsilon_i(t)$. The spectrum of Lyapunov exponents for the trajectory $\mathbf{V}(t)$ is defined as

$$\lambda_i = \lim_{t \to \infty} \lim_{\epsilon(0) \to 0} \frac{1}{t - t_0} \log_2 \frac{\epsilon_i(t)}{\epsilon(0)} \tag{5.6}$$

for $1 \leq i \leq n$. Note that the Lyapunov exponents depend on the trajectory $\mathbf{V}(t)$. Their values are the same for any state on the same trajectory, but may be different for states on different trajectories. The trajectories of an n-dimensional phase space have n Lyapunov exponents and these are often called the *Lyapunov spectrum*. It is conventional to order them according to size. The qualitative features of the asymptotic local stability properties can be summarized by the sign of each Lyapunov exponent: a positive Lyapunov

5.2. LPN FOR DYNAMIC SYSTEM MODELING

exponent indicating an unstable direction, and a negative exponent indicating a stable direction.

In order to obtain an estimate for a particular trajectory, the divergence at many points on the trajectory can be averaged. For example, the characteristic (largest) exponent can be computed:

$$\lambda = \lim_{N \to \infty} \frac{1}{N} \sum_{i=1}^{N} \frac{1}{(t_i - t_0)} \log_2 \frac{\epsilon(t_i)}{\epsilon(t_0)} \quad (5.7)$$

If $\lambda \leq 0$ the motion is regular, whilst if $\lambda > 0$ the motion is chaotic. For all practical purposes the complexity of the system increases with λ, and the choice of Δt in equation (5.3) depends on λ, since the magnitude of the exponent reflects the time scale on which trajectories diverge. In other words, a smaller value of Δt has to be chosen if λ is large for some trajectory of the system.

5.2.1 LPN for a Two Point Attractor System

In this section, the dynamic system (5.4) is considered. The characteristic Lyapunov exponent of the above system was calculated (using the method described in [181]), to be about -0.26. With $p = 1$ and $\Delta t = 1$, a 4-10-2 neural network was trained for 30,000 iterations using backpropagation ($\eta = 0.2, \alpha = 0.1$). The training data consisted of 198 scaled points, from a single trajectory whose initial point was $(x(0), \dot{x}(0)) = (-1, -0.5)$. Thus, the network was constructed to satisfy:

$$(x(t), \dot{x}(t), x(t-1), \dot{x}(t-1)) \longrightarrow (x(t+1), \dot{x}(t+1)), \quad t \in [0, 198] \quad (5.8)$$

for this particular trajectory of the given system (5.4). The crucial question for system modeling is whether the constructed network satisfies the mapping (5.8), for *all trajectories*. Thus, what is interesting is the generalization capability of the LPN.

One way to test such a network, might be to start it at an initial point $(x_0, \dot{x}_0)$, and iterate the network (with no reference to the actual dynamic system), to see how well it follows the desired trajectory. Given that the system might in general be chaotic any slight error in a local prediction will, if uncorrected and iterated, rapidly lead to significant deviations from the true trajectory. An analogy might be that of a standing man suddenly rendered devoid of sensory feedback, who would rapidly fall over. However, the result of such a test is expected to be unsuccessful, since the LPN is constructed to predict only locally in time.

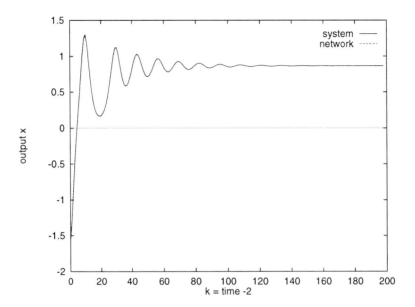

Figure 5.4: Two point attractor system. How well the 4-10-2 LPN learnt the training data. The two graphs are virtual indistinguishable. The network is predicting 1 sec ahead.

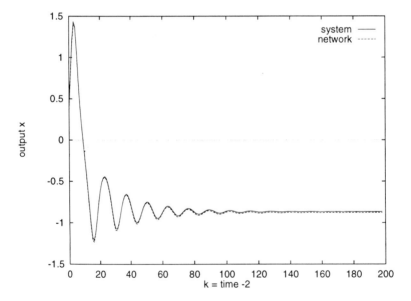

Figure 5.5: Two point attractor system. Generalization capability of the 4-10-2 LPN on a trajectory from initial point (-0.5, 0.5), i.e. different from the training trajectory. Network predicting 1 sec ahead.

5.2. LPN FOR DYNAMIC SYSTEM MODELING

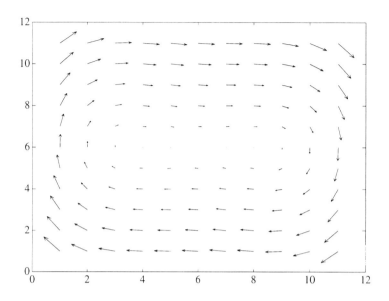

Figure 5.6: The vector field for the two point attractor system.

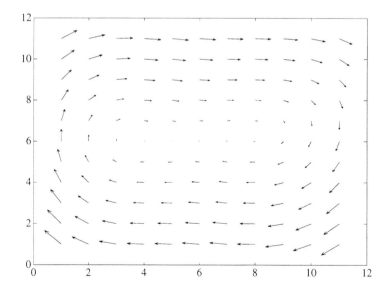

Figure 5.7: Two point attractor system. The vector field of the 4-20-10-2 LPN - network predicting 0.01 sec ahead.

Consequently other ways are sought for examining how much of the global dynamics a LPN can abstract from a limited set of training examples. Therefore the ability of the network to predict one step ahead is examined. The performance of the network on the training trajectory is shown in Figure 5.4, in which we see very good local prediction using the network compared with the trajectory upon which it was trained. Of course, this was to be expected and merely illustrates that the network has converged on the training data. Extending the trajectory also showed that the extrapolation performance of the LPN for the training trajectory was similarly very accurate.

The next step is obviously to test the network on a trajectory different from that upon which it was trained. The result of such a test of *generalization* is shown in Figure 5.5. The trajectory starts at (-0.5, 0.5), $t \in [0, 50]$ and again the agreement is very accurate.

Another way to quickly verify that a LPN can abstract a good approximation to the overall dynamics from observations of a single trajectory is to train a LPN using a relatively small prediction interval Δt. Then the overall differential vector field can be computed, for both the dynamic system and the LPN (for a large Δt this would not be very meaningful) and compare them. This is done in Figure 5.6 and Figure 5.7 respectively. The 4-20-10-2 ($p = 1, \Delta t = 0.01$) network (trained for 50,000 iterations) here requires more hidden nodes, because rather small relative changes must be modeled, but the successful modeling phenomenon is clearly demonstrated.

5.2.2 LPN for the Van der Pol Equation

As another example, consider the Van der Pol equation

$$\ddot{x} - \mu(1 - x^2)\dot{x} + x = 0 \tag{5.9}$$

Again this can be transformed into a system of first order differential equations, so as before the Runge-Kutta method can be used:

$$\dot{x} = y$$
$$\dot{y} = -x + \mu(1 - x^2)y$$

The Van der Pol equation can be regarded as describing a mass-spring-damper system with a position-dependent damping coefficient. For large values of x, the positive damping coefficient removes energy from the system, and thus the motion has a tendency to converge. However, for small values of x, the

5.2. LPN FOR DYNAMIC SYSTEM MODELING

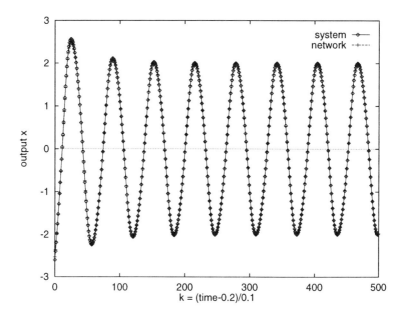

Figure 5.8: Van der Pol system. How well the 4-10-2 LPN learnt the training data. The network is predicting 0.1 sec ahead.

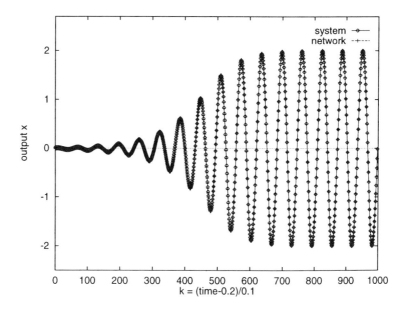

Figure 5.9: Van der Pol system. Generalization capability of the 4-10-2 LPN on trajectory starting at $(0.01, 0.01)$, i.e. different from the training trajectory. The network is predicting 0.1 sec ahead.

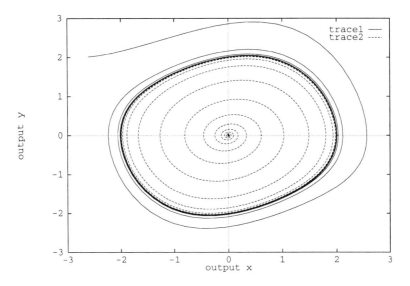

Figure 5.10: Van der Pol system. A phase plane diagram in which two different trajectories are shown: one starting at $(-3, 2)$ and spirals inwards and the other starts near the origin $(0.01, 0.01)$ and spirals outwards.

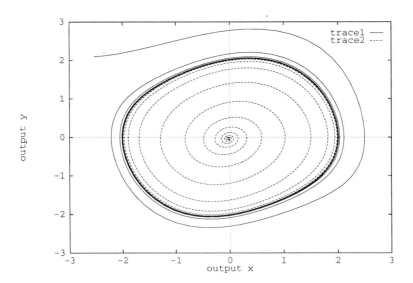

Figure 5.11: Van der Pol system. The phase space diagram for the 4-10-2 LPN for the trajectories starting at $(-3, 2)$ and $(0.01, 0.01)$ respectively. The network is predicting 0.1 sec ahead.

5.2. LPN FOR DYNAMIC SYSTEM MODELING

damping coefficient is negative, and the "damper" adds energy to the system so that the motion has a tendency to diverge. As a result of this nonlinear damping, the system motion can neither grow without bound, nor decay to zero. Instead, it displays a sustained oscillation independent of initial conditions, a *limit cycle*. It is sustained by periodically releasing energy into and absorbing energy from the environment, via the damping term. The characteristic Lyapunov exponent of the above system, calculated using the method in [181], is approximately zero.

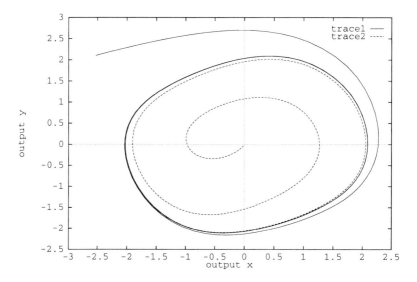

Figure 5.12: Van der Pol system. Iterating the 4-10-2 LPN for the trajectories starting at $(-3, 2)$ and $(0.01, 0.01)$.

Thus the system has a stable periodic solution whose period and amplitude depend on the parameter μ. In these experiments a value of $\mu = 0.2$ was used. Using a number of experiments it was decided that one past state ($p = 1$) of the dynamic system (5.9) should be used for the LPN method with $\Delta t = 0.1$.

Therefore a 4-10-2 LPN was trained for 100,000 iterations (using a learning rate and momentum of $\eta = 0.2, \alpha = 0$ respectively), on 500 points taken from the trajectory through (-3,2). In Figure 5.8 the successful result of the training is shown. In order to test the generalization capability of this LPN, the trajectory that starts at (0.01,0.01) is plotted (Figure 5.9). Again the generalization capability of the network is very accurate for the one-step prediction. Figures 5.10 and 5.11 illustrate phase space diagrams of the actual system and the

LPN respectively, for the two trajectories starting at (-3,2) and (0.01,0.01). It is interesting to see that although the training trajectory is outside the region enclosed by the limit cycle, the network nevertheless performs very well on the trajectory inside the region.

The same network when used for multistep (iterative) prediction gives, as expected, a poor performance but nevertheless it still has a limit cycle (Figure 5.12).

5.2.3 LPN Applied to a Chaotic System

In this section a system which exhibits chaotic behavior is considered:

$$\ddot{x} + 0.1\dot{x} + x^5 = 6\sin t \qquad (5.10)$$

This represents a lightly damped, sinusoidally forced mechanical structure undergoing large elastic deflections. In this case, the system is extremely sensitive to initial conditions, i.e. is chaotic.

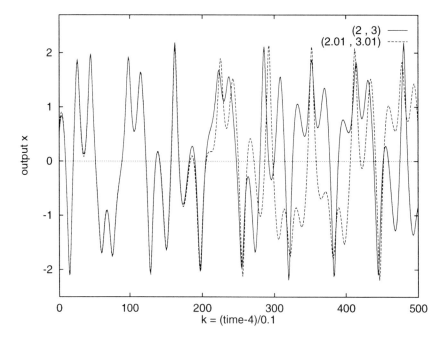

Figure 5.13: Chaotic system. Two trajectories with two almost identical initial conditions.

Like the previous dynamical systems we considered, in order to get all the necessary data to train a LPN, it is transformed (5.10) in a system of first-order

5.2. LPN FOR DYNAMIC SYSTEM MODELING

differential equations:

$$\dot{x} = y$$
$$\dot{y} = -0.1y - x^5 + 6\sin t$$

Figure 5.13 illustrates two trajectories with two almost identical initial con-

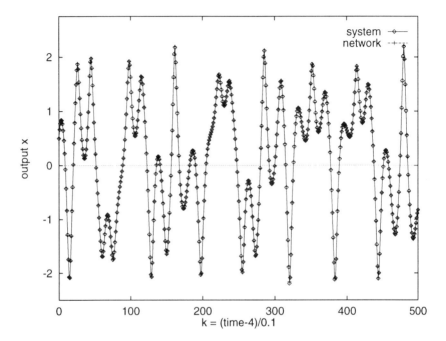

Figure 5.14: Chaotic system. How well the 8-10-5-2 LPN learnt the training data. Network predicting 0.1 sec ahead.

ditions, namely $(x(t), \dot{x}(t)) = (2, 3)$ and $(2.01, 3.01)$ at $t = 0$. The characteristic Lyapunov exponent for this system, calculated using the method in [181], is about 0.238.

The empirical study of (5.10) led to the choice of $p = 3$, $\Delta t = 0.1$ and a 8-10-5-2 architecture for the LPN. The network was trained using a learning rate of $\eta = 0.2$, and momentum $\alpha = 0$ and 500 data points taken from the trajectory through $(2,3)$ for $t \in [0, 50]$ for $100,000$ iterations (see Figure 5.14). The generalization capability of the LPN for the chaotic dynamics was very good and is shown in Figure 5.15 (trajectory through $(2.01, 3.01), t \in [0, 50]$).

How well the LPN models the attractor of the chaotic system (5.10) can be visualized by constructing the Poincaré section (stroboscopic plots at times that are multiples of 2π) of the dynamic system and of the LPN. Starting with

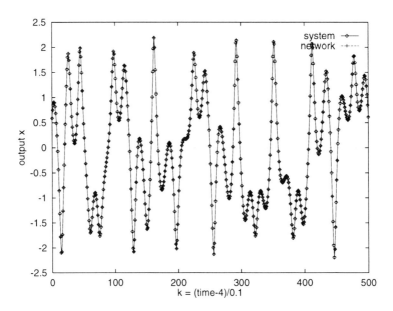

Figure 5.15: Chaotic system. Generalization capability of the 8-10-5-2 LPN on trajectory starting at $(2.01, 3.01)$. Network predicting 0.1 sec ahead.

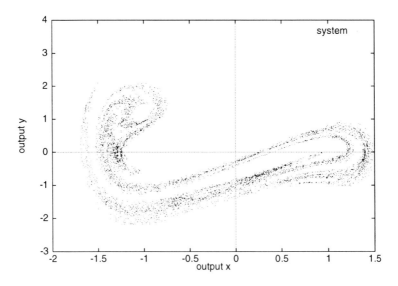

Figure 5.16: Chaotic system. A Poincaré section.

5.2. LPN FOR DYNAMIC SYSTEM MODELING

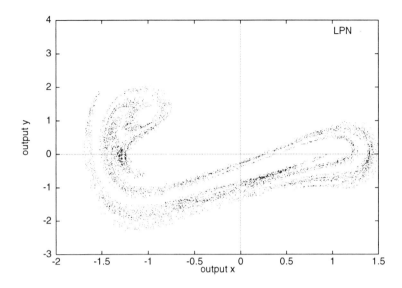

Figure 5.17: Chaotic system. A Poincaré section for the 8-10-5-2 LPN. Network predicting 0.1 sec ahead.

the initial point $(2,3)$ the dynamic system evolution $(x(t), \dot{x}(t))$ was computed for $t \in [0, 40,000]$ in steps of 10^{-5}.

The results are shown in Figure 5.16 and Figure 5.17. It seemed rather striking that a neural network trained on a single trajectory could abstract such a detailed local model of the overall dynamics.

5.2.4 LPN for the Euler Equations

In this section, the system which describes the motion of a rigid body (Euler equations) is considered. As seen in Section 3.2, a certain combination of parameters for linear feedback control lead the system into a chaotic motion. This chaotic system will be the subject of this section. The equations describing it are rewritten (with x, y, z substituting the angular velocities $\omega_1, \omega_2, \omega_3$ about the three principal body axes respectively):

$$\begin{aligned} I_1 \dot{x} &= (I_2 - I_3)yz + L \\ I_2 \dot{y} &= (I_3 - I_1)zx + M \\ I_3 \dot{z} &= (I_1 - I_2)xy + N \end{aligned} \quad (5.11)$$

where $I_1 = 3I_0$, $I_2 = 2I_0$, $I_3 = I_0$, $I_0 = 1$ and

$$\begin{pmatrix} L \\ M \\ N \end{pmatrix} = \begin{pmatrix} -1.2 & 0 & \sqrt{6}/2 \\ 0 & 0.35 & 0 \\ -\sqrt{6} & 0 & -0.4 \end{pmatrix} \begin{pmatrix} x \\ y \\ z \end{pmatrix} \qquad (5.12)$$

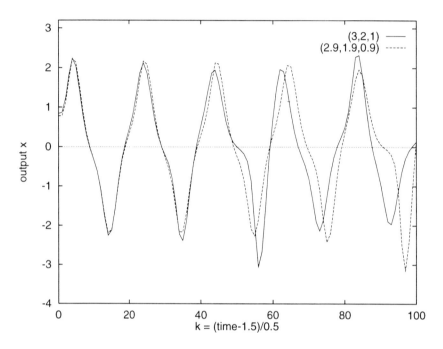

Figure 5.18: Euler system for the chaotic control regime. Two trajectories with two almost identical initial conditions.

Figure 5.18 shows the x angular velocity of the system, for two trajectories with similar initial conditions, $(3, 2, 1)$ and $(2.9, 1.9, 0.9)$. The empirical study of the system lead to the choices of $p = 3$ and $\Delta t = 0.5$, and a 12-10-5-3 LPN was trained for $90,000$ epochs with 501 data on the trajectory through $(3, 2, 1)$. The results of the training are shown in (Figure 5.19). The generalization capability of the network is tested for all three angular velocities, using the trajectory through $(1.9, 1.9, 1.9)$ (Figures 5.20,5.21, 5.22) which is very different from the training trajectory. From all the diagrams can be seen the very accurate one step prediction capability of the network.

Of course, if the trained network is used for iterative prediction, then (as expected), there is a rapid divergence, see Figure 5.23. The performance of the iterated network can be compared with the actual dynamics, by plotting first

5.2. LPN FOR DYNAMIC SYSTEM MODELING

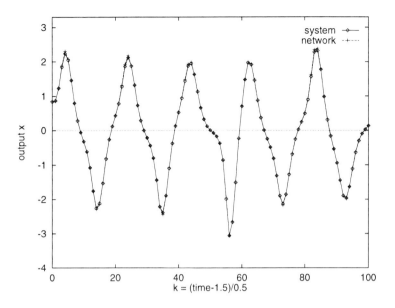

Figure 5.19: Euler system. How well the 12-10-5-3 LPN learnt the training data (only a part of the training trajectory is shown) - x angular velocity. Network predicting 0.5 sec ahead.

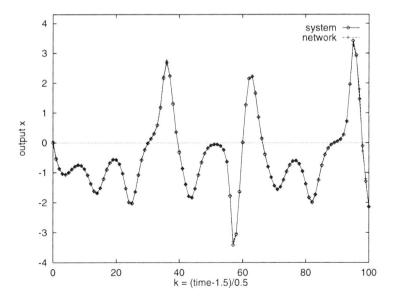

Figure 5.20: Euler system. Generalization capability of the 12-10-5-3 LPN on trajectory starting at $(1.9, 1.9, 1.9)$ - x angular velocity. Network predicting 0.5 sec ahead.

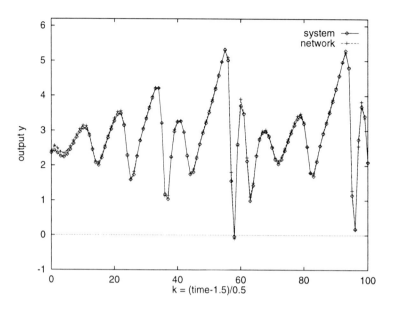

Figure 5.21: Euler system. Generalization capability of the 12-10-5-3 LPN on trajectory starting at $(1.9, 1.9, 1.9)$ - y angular velocity. Network predicting 0.5 sec ahead.

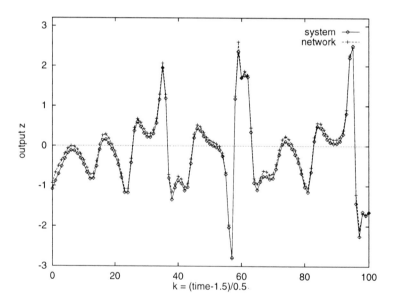

Figure 5.22: Euler system. Generalization capability of the 12-10-5-3 LPN on trajectory starting at $(1.9, 1.9, 1.9)$ - z angular velocity. Network predicting 0.5 sec ahead.

5.3. MODELING AND ADAPTIVE CONTROL

the absolute divergence of the x angular velocity between the system trajectory through $(3, 2, 1)$ and the perturbed system trajectory $(2.9, 1.9, 0.9)$ (Lower trace labeled *system* in Figure 5.24), and second the absolute divergence of the x angular velocity between the system trajectory starting at $(3, 2, 1)$ and the iterated network trajectory starting at $(2.9, 1.9, 0.9)$ (upper trace labeled *network* in Figure 5.24). Whereas the perturbed system begins to diverge significantly after 20 steps, the iterated network on the generalization trajectory is beginning to diverge significantly after 10 steps.

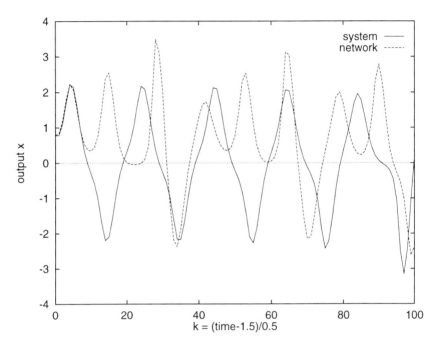

Figure 5.23: Euler system. Iterative prediction of the 12-10-5-3 LPN on trajectory starting at $(2.9, 1.9, 0.9)$ - x angular velocity.

The Poincaré map for the $x - z$ plane of the system is shown in Figure 5.25, and the corresponding diagram for the LPN in Figure 5.26.

5.3 The Importance of Modeling to Adaptive Control

To find a suitable set of control signals for a plant at any given moment it may, in general, be necessary to search the space consisting of potential control inputs

94   CHAPTER 5. NEUROMODELS OF DYNAMIC SYSTEMS

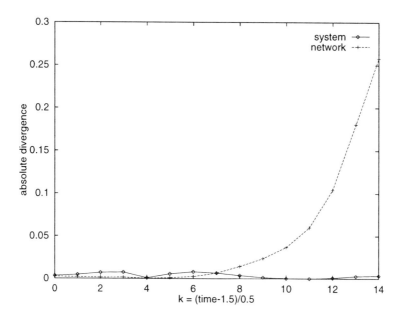

Figure 5.24: Euler system. Divergence comparison for iterative prediction (see text).

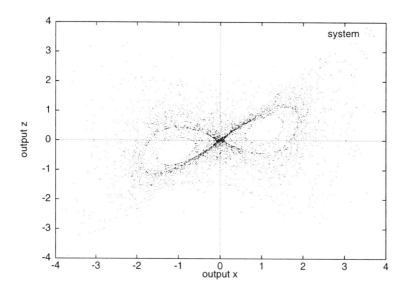

Figure 5.25: Euler system. Poincaré section through the $x - z$ plane for the "real" system. (From *Neural Networks and their Applications*, J.G.Taylor [Ed.]. Copyright John Wiley & Sons Limited. Reproduced with permission.)

5.3. MODELING AND ADAPTIVE CONTROL

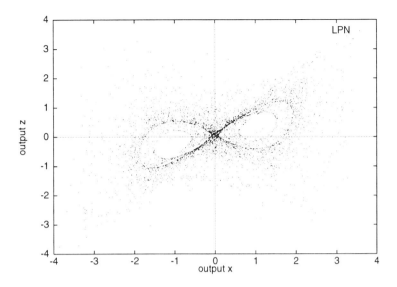

Figure 5.26: Euler system. The corresponding Poincaré section through the $x-z$ plane for the 12-10-5-3 LPN. LPN predicting 0.5 sec ahead. (From *Neural Networks and their Applications, J.G.Taylor [Ed.]*. Copyright John Wiley & Sons Limited. Reproduced with permission.)

and the corresponding system response a short time ahead, and to determine an optimal or near optimal set of control signals. In order to achieve this, we first require a reasonably accurate model which informs us, given the current state, what the system response would be to a hypothetical set of control inputs. In situations where the system dynamics have changed, or are *a priori* unknown, it is first necessary to *learn* such a model. The above results show that we may learn such a dynamic model by constructing a LPN neuromodel whose inputs are the current state and control inputs and whose outputs are the next system state. Moreover, the results suggest that for analytically smooth dynamic systems relatively few training examples are required in order to form a reasonably accurate model over a relatively large region of the space of inputs.

If we assume that such LPN neuromodel can be constructed (as shown), e.g. for the system associated with the attitude control problem, then we may also suppose that it could be implemented in hardware with an extremely rapid throughput. This would imply that in the time between one control input and the next, perhaps a significant fraction of a second, we could compute many responses to possible control inputs and select the most appropriate. This is broadly the approach that will be followed in the next chapters, except that the search will not be random, but based on genetic algorithms (see Chapter 7).

The genetic algorithm implementation can be relatively simple (the control signals are encoded as binary strings) so it is feasible to implement it in discrete logic hardware if necessary. As will be seen in Chapter 8 a genetic algorithm can be very efficient in finding solutions to the inverse kinematic problem.

Chapter 6

Current Neurocontrol Techniques

Besides the electrical engineering theory of the transmission of messages, there is a larger field which includes not only the study of language but the study of messages as a means of controlling machinery and society, the development of computing machines and other such automata, certain reflections upon psychology and the nervous system, and a tentative new theory of scientific method.

Norbert Wiener

(The Human Use of Human Beings)

The use of neural networks in control applications (including process control, robotics and aerospace applications, among others) which is referred to as neurocontrol, has begun a pattern of very rapid growth [209]. Neurocontrol can enable the automated control of systems, which could not be controlled in the past for two reasons: the physical cost of implementing a known control algorithm, or the difficulty of finding such an algorithm for complex, noisy, nonlinear problems. The neural network approach is very attractive because of their ability to learn, to approximate functions, to classify patterns and because of their potential for massively parallel hardware implementation.

Even when a task can be done equally well using conventional methods, or neural nets, there are several advantages to using the latter. For example,

Intel, among others, has produced a neural net chip which has more effective throughput (for neural net calculations) than all of the Crays of the world put together [214]. By translating a method into something which can use this chip (or similar chips), it might be possible to produce a system which, when fully developed, requires only the insertion of a few extra chips into the system to effect control.

Since 1988, many papers have been published on neurocontrol and in general there are four basic approaches: supervised control, direct inverse control, neural MRAC-type adaptive control and reinforcement learning.

6.1 Supervised Control

In supervised control there is a need for a training set of $X(t)$ and $u^*(t)$, where u^* is a vector of targets for the action vector. The target actions usually come from recording the actions of human beings solving the desired control problem.

The motivation behind supervised control is approximately the same as the motivation behind expert systems: to "clone" the skills of a human being (Figure 6.1). However, instead of copying what a person says, we copy what he or she does. Generally, it can be considered as a tool for transferring information from a human expert to a robot. Supervised controllers record both actions and sensor inputs, so that the system can learn how movements vary in response to sensor inputs.

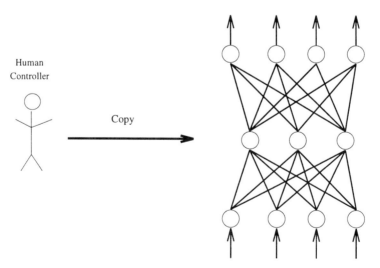

Figure 6.1: The supervised control paradigm.

When it is possible to find a human who can perform the task adequately, supervised control has the advantage of being straightforward and relatively certain of success. If a human is already available, capable of performing the task, neurocontrol may still be worthwhile if the supply of such humans is limited, if the humans are expensive to use (compared with a neural net chip implementation), or if the humans can only control the system safely when it is simulated and slowed down.

The earliest example of a supervised control system was Widrow's broom balancer in the 1960's. Since then, many other supervised control applications have appeared (for example see [141]).

6.2 Direct Inverse Control

Inverse models are very important in many areas, including control. Many controllers use direct inverse models of the plant. Direct inverse control is also based on supervised learning. Once again, any supervised learning method may be used. The targets are the actuator signals, and the inputs are the coordinates describing the state of the plant. The history of inputs and targets comes from letting the system move around, and recording actual states and actuator signals.

Generally, the problem of finding an inverse model of a dynamic system is ill-posed, in the sense that the solution to the problem is not unique. If the dynamic system under consideration has a many-to-one mapping from actions to system state, then there are a number of possible inverse models. A typical problem in direct inverse control associated with ill-posed systems is illustrated in Figure 6.2.

The goal is to build a controller that will have as input a desired location $X(t)$ specified by a higher level controller or a human, and output the control signals, $u(t)$, which will move the robot arm to the desired location. The desired location is in two-dimensional space, and the control signals are simply the joint angles θ_1 and θ_2. If dynamic effects are ignored, then it is reasonable to assume that x_1 and x_2 are a function f, of θ_1 and θ_2 as shown in the figure. This kinematic function is a many-to-one mapping: for every Cartesian position that is inside the boundary of the workspace, there are an infinite number of joint angle configurations which achieve that position. This implies that the inverse kinematic relation $f^{-1}(x)$ is not a function; rather, there are an infinite number of inverse kinematic functions corresponding to particular

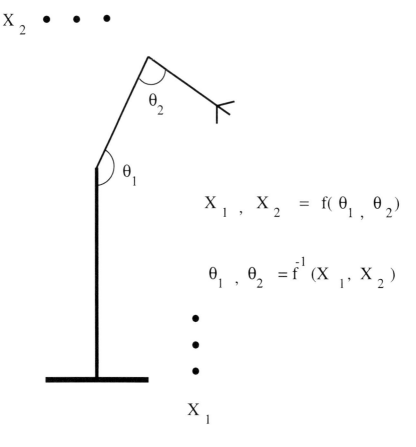

Figure 6.2: Direct inverse control.

choices of points θ_1, θ_2 in the inverse images of each of the Cartesian positions. The problem of learning an inverse kinematic controller for the arm, is that of finding a particular inverse among the many possible mappings.

Most applications of direct inverse control treat the problem as a classical supervised learning problem [223]. The idea is to build a database of inputs-targets by observing the input-output behavior of the environment, and then to use the database as the training set of a supervised technique by reversing the observed inputs-outputs. However, if the plant has a many-to-one mapping, then the network will learn an "incorrect" inverse model of the plant. This is because the training set consists of one-to-many data, and the fact that standard supervised learning algorithms average over multiple targets (assuming a squared-error criterion function)[1]. In addition to the non-uniqueness of a

---
[1] Generally, if any plant state has an inverse image that is a non-convex region in action

6.2. DIRECT INVERSE CONTROL

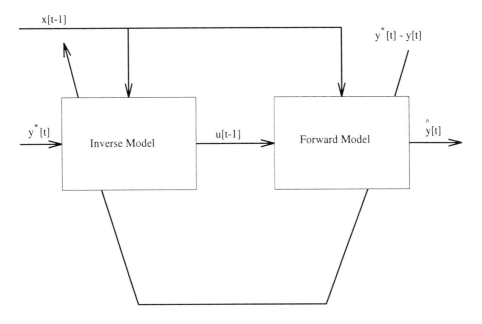

Figure 6.3: Jordan's approach for learning an inverse model.

inverse mapping, this approach is not goal-directed, since in order to construct the training set, one is sampling the action space. Besides that, for certain plants inverse control can introduce instability to the system.

Jordan and Rumelhart [110] used a different type of inverse control, in order to face the problem of non-uniqueness of inverse models. In their approach, they first train a forward model of the plant (able to give an estimation $\hat{y}[t]$ of the actual output $y[t]$ of the plant) and then they train a neural network controller by propagating back the errors $y^*[t] - y[t]$ of the system (where $y^*[t]$ is the desired response) through the forward model (Figure 6.3). In their work, when they deal with plants that are required to follow a particular trajectory, they use a modified algorithm, equivalent to that of backpropagation-through-time [175], [213]. A similar approach was taken by Kawato et al. [113], [114], [115] and D. Psaltis [168]. Generally however, it is not always possible to reach a desired next state from any other dynamic system state, and therefore their methods are not generally applicable to control problems. In particular such an approach cannot be applied directly to the attitude control problem where the goal is to reach a target state starting from an initial state.

---

space, then there exist sets of samples in the region whose average lies outside of the region, therefore yielding an incorrect inverse model.

6.3 Neural MRAC Adaptive Control

Classical model reference adaptive control (MRAC) techniques solve a problem very similar to that of direct inverse control. The user, or higher-order controller is asked to supply a reference model that outputs a desired trajectory. The control system is asked to send control signals to the plant, so as to make the plant follow the desired trajectory. It is usually assumed that the plant and the controller are both linear, but that the parameters of the plant are not known ahead of time. Various techniques have shown how to adapt the parameters of the control system to solve this problem.

Neural MRAC adaptive control has been defined as the use of artificial neural networks in place of more classical mappings (matrices), within the classical designs of adaptive control theory [141]. This approach is mainly represented by the work of Narendra at Yale University.

Narendra has developed three designs so far that extend MRAC principles to neural networks [147]. The first is essentially just direct inverse control, adapted very slightly to allow the desired trajectory to come from a reference model. The second, which he called his most powerful and general in 1990, is very similar to a strategy used also by Jordan. In this approach, the problem of following a trajectory is converted into a problem in optimal control [29], simply by trying to minimize the gap between the desired trajectory and the actual trajectory. This is equivalent to maximizing the following utility function:

$$U = -1/2 \sum_{t,i}(X_i(t) - X_i^*(t))^2 \qquad (6.1)$$

where $X(t)$ is the actual position at time t and X^* is the desired or reference position.

Narendra's third design is similar to the second design, except that radial basis functions are used in place of multilayer perceptrons, and certain constraints are placed on the system. Stability theorems (which require strong assumptions for the plant) have been proved: in practice however, plant characteristics like delays and reversals make these theorems inapplicable [218].

Generally, the published work in neural MRAC adaptive control by Narendra concerns relatively simple discrete systems [141], [149], [150], [151], [152], [153].

6.4 Reinforcement Learning Control

Reinforcement learning is the process by which the response of a system (or organism) to a stimulus is strengthened by reward and weakened by punishment [137]. Rewards and punishments represent favorable and unfavorable environmental reactions (respectively) to a response. Such reactions are evaluated by a system in its effort to achieve its goals. A reinforcement learning system seeks its goals by strengthening some responses and suppressing others.

Reinforcement learning control is based on psychological learning theories [137]. A controller receives evaluation of an action based on the definition of a utility function or reinforcement variable U. If the evaluation is positive (favorable) then the probability associated with that action is increased, and the probabilities associated with all other actions are decreased (Figure 6.4). Conversely, if the evaluation is negative (unfavorable), then the probability of the given action is decreased and the probabilities associated with all other actions are increased. Loosely speaking, the reward signal positively reinforces those states of the controller (network) that contribute to improvement, while the punishment signal negatively reinforces the states that produced improper behavior. Reinforcement control can be viewed as an adaptive optimal control approach [194].

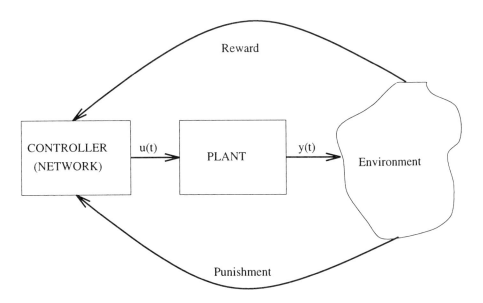

Figure 6.4: The underlying idea of reinforcement learning control methods.

The above characteristic features of reinforcement learning algorithms differ in important ways from the corresponding features of supervised learning algorithms. Supervised learning algorithms are based on the existence of a signed error vector rather than an evaluation.

Many researchers use different names, like adaptive critics, approximate dynamic programming (ADP) [211], [212], or neurodynamic programming [20] to refer to reinforcement learning control. The key point in all of these is the same as that of dynamic programming. Given a utility function U which has to be maximized over all future time (which could be an infinite or a finite time horizon), find a strategic utility function J, for which maximization in the short term will maximize U in the long term [217]. However, exact dynamic programming is too "expensive" computationally for complex dynamic systems or large problems, and this is why the approximate dynamic programming (adaptive critics) techniques are used. A neural network (called the *critic network*) learns to output (predict) an approximation J^* of the function J (Figure 6.5). Thus the Bellman equation in dynamic programming

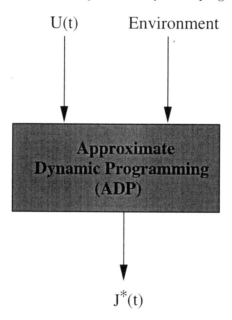

Figure 6.5: The Approximate Dynamic Programming (ADP) operation.

$$J(\mathbf{R}(t)) = \max_{\mathbf{u}(t)} [U((\mathbf{R}(t)), \mathbf{u}(t)) + <J(\mathbf{R}(t+1))>] \qquad (6.2)$$

is not solved exactly, but it is replaced by a neural network able to approximate

6.4. REINFORCEMENT LEARNING CONTROL

$J(\mathbf{R})$ by $J^*(\mathbf{R})$ (this method is also called *heuristic dynamic programming* (HDP)). In (6.2) $\mathbf{R}(t)$ is the state of the environment at time t, $\mathbf{u}(t)$ is the control input vector to the plant, and $<J(\mathbf{R}(t+1))>$ is the expected value of J at time $t+1$.

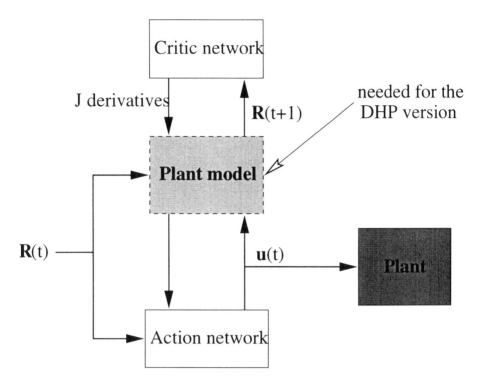

Figure 6.6: The overall operation of HDP, DHP, and GDHP for neurocontrol. The plant model is needed only for the DHP.

Two other variations of approximate dynamic programming are those of *dual heuristic dynamic programming* (DHP) and *globalized dual heuristic dynamic programming* (GDHP). In the DHP, the critic network is a neural network which outputs an approximation of the derivatives of $J(\mathbf{R}(t))$, i.e. $\partial J^*(\mathbf{R}(t))/\partial R_i(t)$. In the GDHP, the critic network predicts an approximation of $J(\mathbf{R}(t))$, so as to minimize the errors in the derivatives of J [220].

In all of the versions of approximate dynamic programming, the critic network provides its output to an *action network* (the actual controller) which is responsible for the derivation of the actions (control inputs) $\mathbf{u}(t)$ to the plant. Figure 6.6 shows the overall operation of approximate dynamic programming used in neurocontrol. In the case of DHP, a model of the plant is needed (another neural network similar to those in Chapter 5), in order to backpropagate the derivatives of $J(\mathbf{R}(t))$ through it.

A whole family of algorithms exists for adaptive critic designs, even though an attempt was made to provide a more unified treatment [166]. Although all of these do not seem to be plausible models of the mammalian brain, Werbos recently argued that three levels of adaptive critic systems are needed to achieve higher order intelligence [218]. Apart from their different functionality, each of these levels should operate at a different speed [217].

In general, different variations of reinforcement control techniques have been successfully used in a number of applications (see for example [141], [167], [220]).

6.5 A General Framework for Adaptive Control

In this section a general framework for adaptive control using neural networks is described (Figure 6.7). The plant dynamics are assumed to be similar to those of a spacecraft or generally any rigid body. It is assumed that the dynamic system has control inputs, referred to as thrusts. In addition it is assumed that no external forces are acting upon the system (except the thrusts).

The initial state of the system is an unstable state, in which there are nonzero angular velocities about the three axes. The primary purpose of the controller is to find the appropriate torques that will bring the body (spacecraft) into a stable state. The stable state is defined as a state in which the angular velocities, about the three principal axes, are zero and the orientation of the spacecraft is as desired. In the following discussion it is shown that this is an *inverse dynamics* problem.

It is supposed that a neural network, labeled the *Online network*, acts as a mapping from a system called *Pilot* to the actual control inputs of the dynamic system. This leads to a highly non-linear response to the *Pilot* inputs. The interface can be trained to produce the correct control signals to, for example, an aircraft for (say) a shallow dive even though the actual control signals required are a highly non-linear function of speed, altitude and current fuel load.

However, suppose that the dynamic system changes its expected characteristics. This could happen in a robot manufacturing system, or an spacecraft, through wear or damage. In this event the interface is no longer appropriate for the actual dynamic system. Thus an adaptive interface is needed.

6.5. A GENERAL FRAMEWORK FOR ADAPTIVE CONTROL

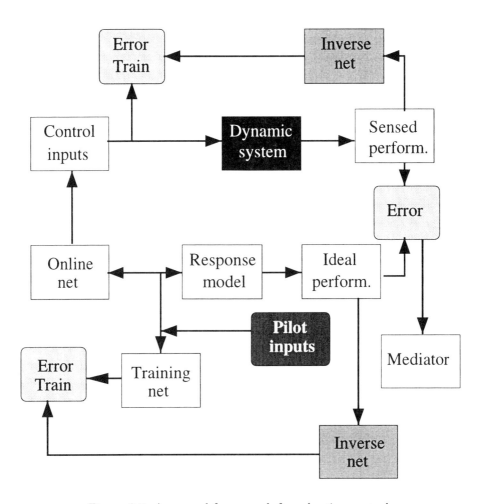

Figure 6.7: A general framework for adaptive control.

What this section suggests is to make the interface adaptive to the currently prevailing dynamics of the actual system using neural networks.

The system must be equipped with sensors to measure all principal parameters and there must be sufficient redundancy in the sensor system in order to provide reliable measurements even if the system as a whole is damaged. The results of the sensor measurements are labeled *Sensed performance* in Figure 6.7.

It can further be supposed that the desired response of the system to any particular *Pilot* signals is known. This is called *Response performance*. The *Ideal performance* takes as inputs the *Pilot* control signals and computes the desired response. If the response is incorrect an *Error* is generated. The existence of an error signal indicates a disparity between the expected response to *Pilot* signals and the desired response (the actual dynamic system has changed). However, with some different *Control inputs* the desired response may still be obtained, or at least a response that more closely approximates the desired response. This indicates an immediate need to modify the interface.

In order to modify the interface, the functional relationship between the *Pilot* signals and the *Sensed performance* has to be changed. In Figure 6.7 this mapping is implemented by a previously trained neural network, the *Online net*. This means that this net will have to be retrained every time the dynamic system changes. Let us now consider how this can be done.

The inputs to the *Online net* are known *Pilot* signals and the outputs are appropriate *Control inputs* to the dynamic system (which is not known). In order to train the new *Online net* the inverse relationship between *Control inputs* and *Sensed performance* is needed, i.e. a functional model from the *Ideal performance* to *Control inputs* which minimizes the overall system *Error* must be calculated. This is an inverse dynamics problem.

In this section a way of solving this adaptive control is described. The *Inverse net* takes as input the *Sensed performance* of the system and tries to output the *Control inputs* which would produce this response. As long as there is no significant difference between the actual *Control inputs* and the outputs produced by the *Inverse net* then it can be considered that the inverse net is functioning correctly. If the system is damaged then it is likely that both the *Inverse net* and the *Online net* will need to be modified. This modification is initiated by the *Mediator* in response to an unacceptable overall system *Error*.

The *Inverse net* is trained on a real time basis. So that after the system is damaged it will adapt to model the actual inverse dynamics of the system. Let

6.5. A GENERAL FRAMEWORK FOR ADAPTIVE CONTROL

us assume that the *Inverse net* has successfully retrained itself. The *Mediator* now uses the *Inverse net* to train the *Training net* on a real time basis. For any given *Pilot* inputs the *Response model* computes the *Ideal performance* and taking this as input the *Inverse net* determines the required *Control inputs*. For the *Training net* the inputs are *Pilot* inputs and the outputs are the *Control inputs* from the *Inverse net*. If permitted the *Mediator* will train the *Training net* over the full space of possible *Pilot* inputs and *Control* inputs. When the discrepancy between the *Pilot inputs* and the *Ideal performance* is lower than it was after the damage occurred the *Training net* and *Online net* are swapped. Now the new *Online net* has a better performance than the previous one and the whole process can begin again with the training net starting from an initial state which was the outcome of the previous round of training.

It was shown that the general problem described above can be reduced to the problem of finding the inverse dynamics of the plant. However, the method described above for learning the inverse dynamics is generally not capable of coping with ill-posed plants. Thus a technique for learning the inverse dynamics of a plant in the general case is required. This can be achieved with the control architecture described in Chapter 8. Then the architecture in Figure 6.7 can be applied to adapt the required interface when necessary.

Chapter 7

Genetic Algorithms

From the war of nature, from famine and death, the most exalted object which we are capable of conceiving, namely, the production of the higher animals, directly follows. There is grandeur in this view of life, with its several powers, having been originally breathed into new forms or into one; and that, whilst this planet has gone cycling on according to the fixed law of gravity, from so simple a beginning endless forms most beautiful and most wonderful have been, and are being evolved.

Charles Darwin
(The Origin of Species)

In nature, the evolutionary process occurs when

- there is a population of self-reproducing entities.
- there is variety among the self-reproducing entities.
- the capability of surviving in the environment is associated with individual variety.

Variety is viewed as a variation in the chromosomes of the entities of the population, something which leads to both the structural and behavioral variation of the entities in their environment.

Entities which perform better at specific tasks (fitter individuals) survive and reproduce at higher rates. This is the concept of *natural selection* introduced by Darwin in 1859 [47]. Because of natural selection, after a number of generations and over a period of time, the population contains individuals that perform their tasks better in the environment. Evolution comes as a consequence of the natural selection scheme.

7.1 Genetic Algorithms

Genetic Algorithms (GAs) first introduced by John Holland [97], are becoming an important tool in machine learning and function optimization mostly, but not necessarily, for situations where it is difficult or impossible to model the external circumstances faced by the problem. Genetic algorithms are based on Darwinian natural selection, evolutionary processes, and natural genetics.

To solve a learning task, a design task, or an optimization task, a genetic algorithm maintains a population of structures (organisms). The probabilistic transformation of this population, through a series of generations according to genetic operators, searches for individuals which represent optimal or near-optimal solutions to the problem.

An individual in the population can represent a whole candidate solution to the problem (Pittsburgh approach) or a part of the solution (Michigan approach). In the latter, the solution is represented by a set of chromosomes in the population [98].

The genetic operators often used are reproduction, crossover and mutation although more advanced genetic operators have been used as well (see [48], [73], [77], and [222]). To apply a genetic operator to one or more individuals of the population, a specific representation of the individuals must be defined [78]. In most applications, this is a string (binary or not), but other representations (e.g. trees) can be used [107] (as in the related field of genetic programming-see Section 7.4).

Reproduction is the operator by which an individual is copied to the next generation (replacing another individual) according to its fitness, defined by a cost or utility function f. Depending on the problem, the function f can be a measure of utility, profit or cost, specified by the problem objectives which must be achieved. According to the natural selection scheme, individuals which have a higher utility (fitness), given by function f, will reproduce at a higher rate than the others.

Crossover between two individuals is defined as the exchange of genes between chromosomes, in order to maximize the fitness of a created offspring. For a binary string individual representation, a simple one-point crossover is

7.1. GENETIC ALGORITHMS

shown in Figure 7.1. A position k is chosen randomly among the length of the string (in the example of Figure 7.1 $k = 5$, if it is assumed that binary bits are numbered from left to right). Two new strings (children) are created by swapping the characters (genes) of the two initial individuals between positions 0 amd k. More sophisticated crossover operators can be defined [49], [73], [170], but the underlying methods are similar.

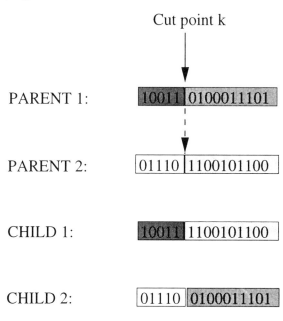

Figure 7.1: A one-point crossover operation.

There are cases (depending on the individual representation) where this simple crossover operation will lead to one or two invalid offspring. For example, in the well known "Traveling Salesman problem" (TSP), one has to find the shortest path for a salesman who would like to visit n cities, starting his tour from city C_1 and ending it in city C_1. This problem is equivalent to finding the shortest cycle in a complete graph of n nodes and it known to be NP-hard. If the adopted representation for the city tour is $C_1C_2 \ldots C_n$ for a solution where the first visited city is C_1, the second is C_2, etc., obviously the previously described crossover operator will create "illegal" children.

To overcome such a problem in the application of a genetic algorithm, one has three choices:

1. apply "repair" operators [49], so as to transform the newly created individuals in a valid form.

2. cancel the crossover operation and try again (this might not be always possible, e.g. when there are too many cancellations, as in the TSP problem, evolution proceeds extremely slowly).

3. go back and redesign the GA, by changing the representation of the individuals so that the normal crossover operation will be applicable without any possibility of invalid children.

Mutation is the occasional (with small probability) random alteration of a gene (Figure 7.2). Although reproduction and crossover effectively search the space by combining the useful features of various individuals, sometimes genetic material is lost (some useful genes are lost). By introducing mutation in the genetic algorithms, this material may be reintroduced to the chromosomes.

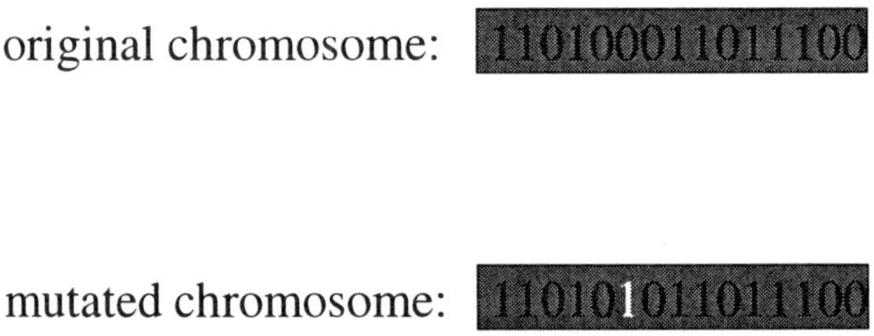

original chromosome: 110100011011100

mutated chromosome: 110101011011100

Figure 7.2: The mutation genetic operation.

Associated with the three main genetic operators of the genetic algorithms, reproduction, crossover and mutation, are the probabilities P_r, P_c and P_m respectively. Thus these operators are applied to the various chromosomes of the population, probabilistically. There are also more sophisticated versions of genetic algorithms where the "correct" operators are enabled by the individuals as a part of the evolutionary process [109]. This dynamic adjustment of bias of the genetic search can affect convergence significantly.

The description of a conventional genetic algorithm is outlined in Figure 7.3. An initial population of size M is chosen randomly in the beginning. Then an iterative process starts until the termination criteria have been satisfied. After the evaluation of each individual fitness in the population, a genetic operator is chosen based on the probabilities associated with each genetic operator. The selected operator is applied to individual(s), chosen with a probability based on fitness. The newly created individuals are inserted into the new population and the loop restarts. The result of a genetic algorithm run is the best individual that appeared through the generations.

7.1. GENETIC ALGORITHMS

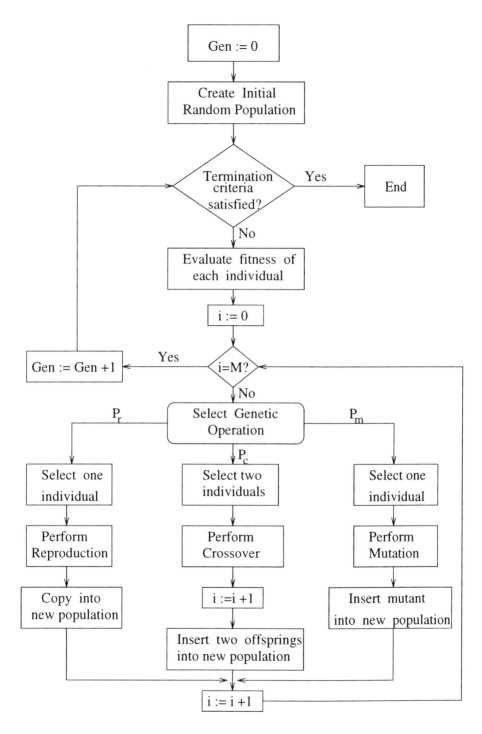

Figure 7.3: The conventional genetic algorithm.

Although the flowchart in Figure 7.3 suggests that in the transition from one generation to another, all the population is replaced, variations of this scheme have proved to improve convergence. The *steady state* model of the genetic algorithm replaces only one or two individuals in the population, so that at the next selection a "parent" is chosen from a competition in the new population [97], [196], [221]. An analysis of the performance of the steady-state genetic algorithm can be found in [197].

7.2 The Genetic Algorithm Underlying Idea

Genetic algorithms do not use previous knowledge about the specific problems which they are trying to solve. In this sense, they are considered "domain-independent". Genetic algorithms search the space of possible individuals (strings) and seek to find high-fitness strings. Viewing the genetic algorithms as optimization techniques, they belong to the class of zero-order optimization methods [205]. According to this, they only require function evaluations instead of additional function derivative information. Therefore, if no derivative information is available, (e.g. in control problems we might deal with a completely unknown plant—the very general case of adaptive control), classical optimization theory is not of much use, but it is still possible to use a genetic algorithm.

Genetic algorithms base their search on the fact that the crossover operations act so as to combine building blocks of good solutions, called *schemata*, from diverse chromosomes. Schemata are just subsets of the space of all possible individuals specified in a particular way. If the individual representation is a binary string, then each building block (schema) is a list of characters belonging to $\{0, 1, *\}$. In order for an individual to contain a certain schema (for example the 01*11*0), it must match the schema in the positions where the schema has a 0, or 1, while it can have any value in the positions where the schema has a * (for the example above, it must match the schema in positions $1, 2, 4, 5$ and 7, where position 1 is the leftmost bit). The * is considered a "don't care" character.

According to the above, the schema is a similarity template, describing a set of strings with similarities at certain positions. John Holland, the founder of the genetic algorithms, came to the conclusion that genetic algorithms base their success on the manipulation of schemata during running, instead of the manipulation of strings. His *Schema Theorem* proved exactly this. Specifically,

7.2. THE GENETIC ALGORITHM UNDERLYING IDEA

the Schema theorem states the following:

Theorem 7.1 (Schema theorem) *A particular schema H receives an expected number of copies in the next generation under reproduction, crossover and mutation, as given by the following equation:*

$$m(H, t+1) \geq m(H,t) \cdot \frac{f(H)}{\bar{f}} \left[1 - p_c \frac{\delta(H)}{l-1} - o(H)p_m \right] \quad (7.1)$$

where $m(H,t)$ is the expected number is the number of schema H in time t, $f(H)$ is the average fitness of the strings representing schema H at time t, $\bar{f}$ is the average fitness of strings in the population, p_c is the crossover probability, p_m is the mutation probability, $\delta(H)$ is the length of a schema defined as the the distance between the first and last specific schema position, l is the string length, and $o(H)$ is the order of a schema defined as the number of fixed positions of the schema.

The Schema Theorem shows that the Darwinian operation of fitness proportionate reproduction, crossover and mutation, causes the number of above average schemata represented in the population to grow from generation to generation. This feature is known as *intrinsic parallelism* [79] because the genetic algorithm is manipulating the number of schema in parallel.

The genetic algorithm paradigm is a search through the space of strings, in actual fact the process yields an intrinsically parallel search through the much larger space of schemata. The main advantages using the GA adaptive strategy are [108]:

- It concentrates samples increasingly towards schemata that contain structures of above average utility.

- Since it works over a knowledge base (i.e. the population of structures) that is distributed over the search space, it is all but immune to getting trapped on local minima (provided the population is sufficiently large).

For the above reasons genetic algorithms can lead rapidly to solutions in complex and multivariable environments.

For the successful application of a genetic algorithm one has to keep in mind that there must always be sufficient diversity in the population. Once the chromosomes in the population become similar to each other, no further evolution is possible (apart from the small changes in the individuals due to the mutation genetic operator). This phenomenon is called *premature convergence* [73].

As mentioned in the previous section the application of a genetic algorithm does not requires knowledge about the problem domain, but requires the availability of some measure of utility or fitness. The initialization of the population is random and fit solutions are "produced" as part of the natural selection and evolution. However, when there is some knowledge about the problem, there is no reason why one should not take advantage of that to improve the convergence of the algorithm and the quality of solutions. This specific knowledge can be used in three ways:

- incorporate it in the fitness function.

- design new, or modify existing, genetic operators to utilize this information.

- initialize the population with structures of medium or low fitness which is higher than that of random initialization. Such an initialization could also come from problem expertise, or even solutions found by another technique.

7.3 Classifier Systems

Genetic algorithms are often used in machine learning tasks in the form of *classifier systems*. Broadly speaking, a classifier system is a system whose interaction with the environment creates a chain of "internal" events (classifiers) of the form

$$if \text{ condition } then \text{ activate} \tag{7.2}$$

which implicitly triggers a chain of external events.

Figure 7.4 shows a typical simple classifier system. A set of sensors detects information flowing into the system from the external environment. This information depends on the previous outputs of the classifier system. The sensors are able to decode the input to one or more messages, which are posted to the message list.

The messages in the message list can activate the classifiers which have the form of a string rule. If a message matches the first part of a classifier (i.e. the part before the colon : in Figure 7.4), the classifier is activated and posts a new message in the message list. For example, given the classifiers in Figure 7.4, if a message 011 arrives in the message list, it will activate the classifier

7.4. GENETIC PROGRAMMING

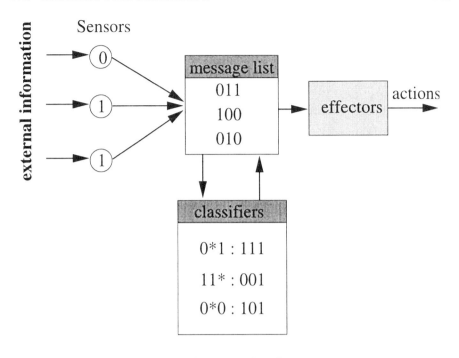

Figure 7.4: A simple classifier system.

$0*1 : 111$ which will post the new message 111 in the message list. Certain messages can trigger an effector in the effector system (Figure 7.4) which will cause an external action to be directed towards the environment.

Note that at any instant, more than one classifier can be activated (as a message can be matched by any number of them). In this case, a competition among the classifiers takes place, in order to designate a winner. The algorithm behind this is known as the *bucket brigade* algorithm [97].

From time to time, new rules (classifiers) have to be derived (learned) and this is done by means of a genetic algorithm. More details about classifier systems can be found in [73].

7.4 Genetic Programming

Genetic Programming (GP)[120], [121] is an extension of genetic algorithms aiming at the automatic discovery of computer programs. The principles of the genetic programming approach are identical to those of genetic algorithms. GP allows a more powerful representation than genetic algorithms, by maintaining a population of structures each of which is a computer program. The characteristics of computer programs (like hierarchical operations, alternative computations based on conditions, iterative calculations, manipulation of different data types and procedural computations) are present in each of the structures in the population.

The representation of a computer program is done through the use of trees with ordered branches. For example in Figure 7.5, the tree shown represents

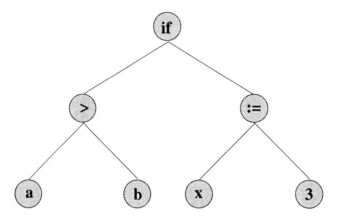

if a>b then x := 3;

Figure 7.5: In genetic programming, the representation of a computer program is a tree.

the single program statement

$$if \quad a > b \quad then \quad x := 3;$$

The second subtree of the root node if is executed (i.e. $x := 3$) only if the evaluation of the first subtree (i.e. $a > b$) is true. This tree representation is equivalent to the parse trees which compilers construct internally to represent a given program.

Like genetic algorithms, the search for a computer program able to solve a particular problem (i.e. the search for a highly fit computer program) is not blind but it is guided by natural selection in an intelligent way. The flowchart for the application of genetic programming also follows that of Figure 7.3. The genetic operators commonly used are: reproduction, crossover and mutation.

The description of the crossover and the mutation operators for genetic programming are shown in Figures 7.6 and 7.7 respectively. Crossover is per-

7.4. GENETIC PROGRAMMING

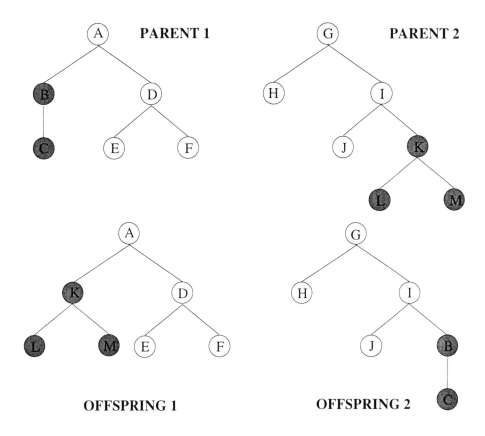

Figure 7.6: The crossover operator in genetic programming.

formed by randomly choosing two points in the chosen parents and swapping the parent subtrees which are defined by these points (Figure 7.6). Note that, unlike genetic algorithms, the two crossover points will be different in general, something which contributes further to the different shapes and sizes of the trees (programs) in the population. Thus there is sufficient diversity in population and the phenomenon of premature convergence is unlikely to happen. Mutation is performed by randomly choosing a node in the chosen individual and by replacing the subtree which starts from this node with a randomly generated tree (Figure 7.7).

A tree in the population is composed of functions and terminals chosen from the function set F and the terminal set T respectively. The functions in the function set may include arithmetic or Boolean operations, conditional operators (such as *if-then-else*), iteration functions (e.g. *while*), operators creating statements similar to those found in programming languages, and any

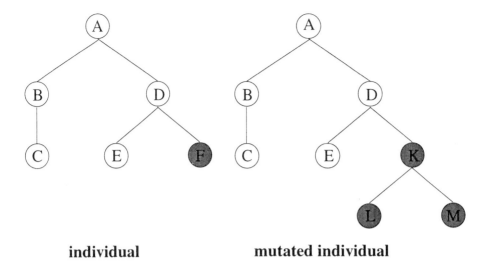

Figure 7.7: The mutation operator in genetic programming.

functions which are closely associated with the problem domain. The terminal set consists of either state variables and inputs or constants (numbers, boolean constants or the random generator R which returns a random value in a pre-specified range).

The function and terminal sets should satisfy the *closure* and *sufficiency* properties. The closure property requires that any function in the function set accepts as an argument any terminal from the terminal set, and any value returned by any other function. Thus, each function should be closed for any combination of arguments that it may receive. For example, the normal arithmetic operator of division always returns a number, unless there is division by zero. To be consistent with the closure property the division function has to be redefined so as to return an "acceptable" value when the denominator is zero. An example of such a definition is shown in Section 7.5. The sufficiency property requires that the sets of functions and terminals which are provided to genetic programming be capable of expressing a solution to the problem under consideration. A solution cannot be expected to be found if the functions or terminals provided are "incomplete".

An important issue in the application of genetic programming is the creation of the initial population. This is usually done randomly, allowing a maximum initial depth of each individual (tree). The depth is defined as the length of the longest nonbacktracking path from the root to a leaf. Research has indicated that the initial population generative method of *ramped half-and-half* gives satisfactory results as it creates enough diversity in the initial population. According to this an equal number of individuals for each depth (up to the maximum) is created. Half of them are full trees (trees for which the length of every nonbacktracking path between the root and any leaf is equal to the prespecified maximum depth) and half of them are of variable shape.

To obtain a solution to a problem one might need to run genetic programming multiple times, (with different initial conditions, i.e. initial population), as the starting point in a search can lead to different results. This is also true for the application of genetic algorithms.

Genetic programming offers an ideal candidate for controller designs, due to the direct matching of its individual structures and the candidate control rules [61]. An example of its application to a part of the attitude control problem (detumbling but not reorientation) is described in Section 7.5.

7.5 Genetic Programming for the Detumbling of a Rigid Body Satellite

A plain vanilla GP was applied for the detumbling of a satellite with moments of inertia $I_x = 23000, I_y = 23300$ and $I_z = 24000$. The goal was to discover a control law which detumbles the satellite about the x, y, z body axes, i.e. $\omega_1 = \omega_2 = \omega_3 = 0$. The dynamics of the satellite are described by equation (3.24), in Chapter 3.

The fitness function used in the GP run was:

$$F = \omega_1^2 + \omega_2^2 + \omega_3^2 \qquad (7.3)$$

with the following terminal and function sets:

Terminal set: $T = (\boldsymbol{\omega}, \dot{\boldsymbol{\omega}}, I, R, e)$

Function set: $F = (\sin, \cos, +, -, *, /, \text{pow})$

where R is the random number generator, / is the *protected division* as defined in [120] (returns 1 if the denominator is 0), and pow is the power function taking as arguments a base and an exponent. The $*$ operator is the standard multiplication function when applied between two numbers, or a number and vector, but is defined as an element-by-element multiplication when both of its arguments are vectors. In this very first approach, the angular velocity $\boldsymbol{\omega}$, the angular acceleration $\dot{\boldsymbol{\omega}}$ and the moments of inertia $\mathbf{I}$ were treated as vectors. If an acceptable solution was not found, then the two vectors would be expanded so that the terminal set would contain the individual elements of $\boldsymbol{\omega}$, $\dot{\boldsymbol{\omega}}$ and $\mathbf{I}$, i.e. $\omega_1, \omega_2, \omega_3, \dot{\omega}_1, \dot{\omega}_2, \dot{\omega}_3, I_x, I_y, I_z$.

The genetic programming parameters shown in Table 1 were used for the results mentioned in this section:

population size	1000
crossover probability	0.90
reproduction probability	0.10
mutation probability	0.00
P_{cf} probability of choosing a function node	0.90
max depth of individuals in initial population	6
max depth of individuals in evolved population	17
generative method	ramped half-and-half
selection method	fitness proportionate

Table 7.1: The genetic programming parameters for the satellite detumbling control problem.

where P_{cf} is the probability of choosing a function node for crossover.

In one of the runs, the following solution was found in generation 61:

$$\mathbf{G} = -1.2 * \mathbf{I} * \boldsymbol{\omega} \tag{7.4}$$

According to the definition of the $*$ operation above, the solution (7.4) can be expanded to the following:

$$\begin{aligned} L &= -1.2 \cdot I_x \omega_1 \\ M &= -1.2 \cdot I_y \omega_2 \\ N &= -1.2 \cdot I_z \omega_3 \end{aligned} \tag{7.5}$$

The control law (7.5) was applied to the satellite (using computer simulations) and the results are shown in Figures 7.8a-c.

Although simulations seemed to show that the control law stabilizes the system for different initial conditions, it is important to try to prove theoretically whether the whole system is stable or not.

Consider the kinetic energy E of the system. This is:

$$E = \frac{1}{2}(I_x \omega_1^2 + I_y \omega_2^2 + I_z \omega_3^2) \tag{7.6}$$

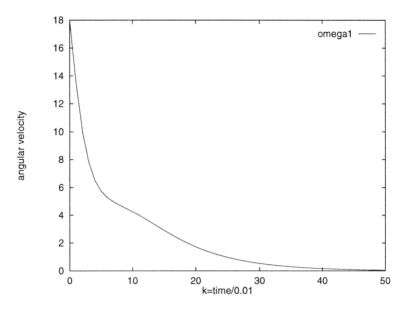

Figure 7.8: The results of the control law discovered by GP applied to the detumbling of the satellite. The satellite moments of inertia are $I_x = 23000, I_y = 23300$ and $I_z = 24000$ and the initial conditions $(\omega_1, \omega_2, \omega_3) = (18.0, 9.0, 9.4)$. The angular velocity ω_1 is shown.

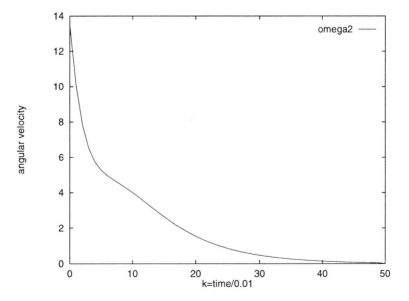

Figure 7.9: The results of the control law discovered by GP applied to the detumbling of the satellite. The angular velocity ω_2 is shown.

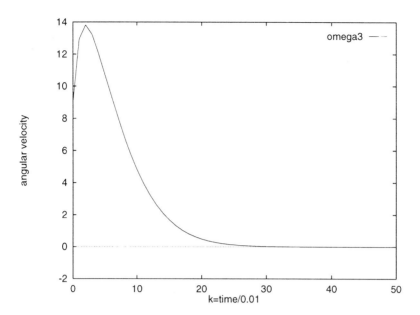

Figure 7.10: The results of the control law discovered by GP applied to the detumbling of the satellite. The satellite moments of inertia are $I_x = 23000, I_y = 23300$ and $I_z = 24000$ and the initial conditions $(\omega_1, \omega_2, \omega_3) = (18.0, 9.0, 9.4)$. The angular velocity ω_3 is shown.

Then
$$\dot{E} = L\omega_1 + M\omega_2 + N\omega_3 \tag{7.7}$$

in a first approximation. If the satellite has to come to rest, then $E \to 0$ and $\dot{E} < 0$. Therefore, applying the constraint

$$\dot{E} = -2.4E \tag{7.8}$$

(where the constant 2.4 could be replaced by any other positive constant–however, different constant values affect the "speed" at which the system evolves) leads to

$$L\omega_1 + M\omega_2 + N\omega_3 = -1.2 \cdot (I_x\omega_1^2 + I_y\omega_2^2 + I_z\omega_3^2) \tag{7.9}$$

with solutions given in (7.5).

Thus the kinetic energy E with the control law (7.5), serves as a Lyapunov function for the system. The highly nonlinear system with the linear control behaves somewhat like a linear system [40].

It can be seen that the control law discovered by the GP approach is stable.

Although previous GP work has been done for control problems, in all of the cases no theoretical stability was proved for those systems. In this section it is shown that sometimes it is possible to prove the theoretical stability of control algorithms found by GP, utilizing the classical theory of Lyapunov functions.

Current research [60] concentrates on the attitude control problem as a whole (detumbling + reorienting a satellite) and whether a GP can find a stable control rule for this problem. Further research has to be done, in various control applications, in order to demonstrate the feasibility of GP for the discovery of stable control regimes.

7.6 The Design of a Genetic Algorithm for the Attitude Control Problem

Assume that the task of a genetic algorithm is the minimization of the following function:

$$U(t) = |\dot{\Theta} + \Theta| + |\Theta| + |\dot{\Phi} + \Phi| + |\Phi| \qquad (7.10)$$

where $\dot{\Theta}, \Theta, \dot{\Phi}, \Phi$ are defined in equations (3.24), (3.32) rewritten here:

$$\begin{aligned} \dot{\Theta} &= \omega_2 \cos \Phi - \omega_3 \sin \Phi \\ \dot{\Phi} &= \omega_1 + \omega_2 \sin \Phi \tan \Theta + \omega_3 \cos \Phi \tan \Theta \\ \dot{\Psi} &= (\omega_2 \sin \Phi + \omega_3 \cos \Phi) \sec \Theta \end{aligned} \qquad (7.11)$$

and

$$\begin{aligned} L &= I_x \dot{\omega}_1 - (I_y - I_z)\omega_2\omega_3 \\ M &= I_y \dot{\omega}_2 - (I_z - I_x)\omega_3\omega_1 \\ N &= I_z \dot{\omega}_3 - (I_x - I_y)\omega_1\omega_2 \end{aligned} \qquad (7.12)$$

As it was shown in Chapter 3 these equations describe the attitude orientation and rotational motion of a rigid body, where L, M, N are the control inputs about the x, y, z principal body axes respectively.

At each time instant t it is desirable to choose the control inputs L, M, N so as to minimize the function F at time $t + \Delta t$. Each control input can take integer values in the range $[-2^{15}, 2^{15}]$.

For the representation of the individuals (chromosomes), binary strings are chosen. The length of each chromosome is 48 bits, with 16 bits dedicated to each of the control variables L, M, N respectively (Figure 7.11).

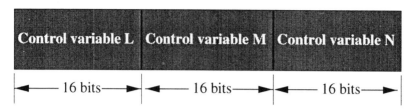

Figure 7.11: Chromosome representation of the genetic algorithm for the attitude control problem. (From *Neural Networks and their Applications*, J.G.Taylor [Ed.]. Copyright John Wiley & Sons Limited. Reproduced with permission.)

The genetic algorithm tries to minimize (7.10) by maximizing the following adjusted fitness function:

$$E(t) = \frac{1}{1 + U(t)} \qquad (7.13)$$

Although, different fitness functions, based on an objective function, have been used in the genetic algorithm literature, all the simulations in this book showed that the choice of (7.13) always leads to good solutions (as long as the other parameters of population size, individual size, etc. are chosen carefully), while others, like the raw fitness function do not. As it was noticed in the book simulations, the adjusted fitness has the benefit of exaggerating the importance of small differences in the value of the standard raw fitness usually used, as the standard fitness approaches 0 (as often occurs on later generations of a run). Thus, as the population of control inputs improves, greater emphasis is placed on the small differences that make the difference between a good control individual and a very good one. This exaggeration is especially important if the standard raw fitness actually reaches 0, when a perfect solution to the problem is found.

The crossover probability was chosen to be $P_c = 0.6$ while the mutation probability was chosen to be $P_m = 0.0333$ [80]. Following the 1-point crossover application to the individuals, mutation with probability P_m is applied to the genes (bits) of each individual. For this specialized genetic algorithm, the two best individuals of the current generation are copied (survive) to the next generation as well. Thus, the very best individuals are not lost and the best

7.6. A GENETIC ALGORITHM FOR ATTITUDE CONTROL

solution found through the generations is the best individual found in the last generation.

A population size of 50 was used and the algorithm terminates after running for 100 generations. The pseudocode of this genetic algorithm for the attitude control problem is shown in Figure 7.12:

$population\_size := 50$;
$generation := 1$;
Initialize population with random binary strings;
while $generation \leq 100$ **do**
    Find the two best individuals of current population;
    Copy the two best individuals to new population;
    for $i = 1$ **to** $population\_size - 2$ **step** 2 **do**
        Select two individuals based on fitness;
        Probabilistically mutate the two individuals;
        Probabilistically perform crossover;
        if $crossover\_performed$ **then**
            Copy the two offspring into new population;
        else
            Copy the two (mutated) individuals into new population;
        endif
    endfor
    $generation := generation + 1$;
endwhile

Figure 7.12: Pseudocode of the genetic algorithm for the attitude control problem. (From *Neural Networks and their Applications, J.G.Taylor [Ed.]*. Copyright John Wiley & Sons Limited. Reproduced with permission.)

An initial population of size 50 is chosen randomly. A loop starts for a 100 generations. During each iteration of this loop, a new population is generated with 50 new individuals. These 50 new individuals consist of the two best of the previous population, and 48 other individuals that are produced with the probabilistical application of the 1-point crossover and mutation. Each pair of individuals is selected for crossover according to the algorithm (based on fitness) which is described in Figure 7.13. After the creation of 50 new individuals which make up the new generation, the termination criterion is checked, and if not satisfied the outer loop restarts.

```
Select()
    j := 1;  sector := 0;
    number := random();
    while sector ≤ number and j ≤ population_size do
        sector := sector + fitness_of_individual_j;
        j := j + 1;
    endwhile
    return individual_{j-1};
end
```

Figure 7.13: Select procedure of the genetic algorithm for the attitude control problem. (From *Neural Networks and their Applications*, J.G.Taylor [Ed.]. Copyright John Wiley & Sons Limited. Reproduced with permission.)

In Figure 7.14 the evolution of the solution to the problem in 100 generations is shown. In this case, the initial conditions (at time $t = 0$) are $\omega_1 = 1.75, \omega_2 = 2.9, \omega_3 = 1.33$ and $\Theta = 1, \Phi = 2, \Psi = 3$, while $I_x = 1160, I_y = 23300, I_z = 24000$. Although in this particular case a minimum was found in 50 generations, in the general case (as the experiments showed) a maximum of 100 generations are required.

The number of possible individuals in the attitude control problem is $2^{16} \times 2^{16} \times 2^{16} = 2.81 \times 10^{14}$. The genetic algorithm described above processes only $48 \times 100 + 2 = 4802$ individuals at each time instant, i.e 1.7×10^{-9} % of the possible control combinations. However, in all of the simulation results run in this book (including these in Chapter 8) the genetic algorithm is able to find a good solution[1] to the problem while processing such a small number of chromosomes. This point is very important for the real time application of the algorithm to the adaptive control architecture described in Chapter 8.

---

[1] Direct estimation of the effectiveness of this genetic algorithm is rather difficult because the true optimum is unknown. In principle it is possible to obtain an upper bound for the achievable fitness from energy considerations; given the fact that available thruster torques are bounded (and no external forces are acting to the body) only states in a certain region about the initial state are in fact reachable. However, the true test of the genetic algorithm is whether it produces sufficiently good solutions to achieve successful control and, as shown in the next chapter, by this test the genetic algorithm is successful.

7.6. A GENETIC ALGORITHM FOR ATTITUDE CONTROL

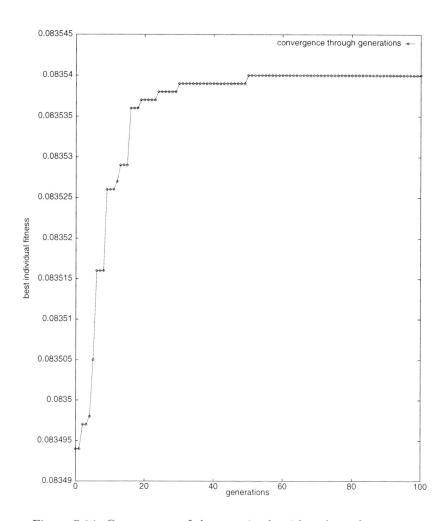

Figure 7.14: Convergence of the genetic algorithm through generations.

Chapter 8

Adaptive Control Architecture

Instead of chasing after the beasts, which would have accomplished little or nothing, Klapaucius, a true theoretician, approached the problem methodically; in squares and promenades, in barns and hostels he placed probabilistic battery-run dragon dampers, and in no time at all the beasts were extremely rare.
Stanislaw Lem
(*The Cyberiad*)

8.1 Introduction

The limitations of the adaptive control techniques that have been described so far, the first encouraging results on simple control problems using neural networks reported in the literature [7, 17, 141], the possibility of hardware implementation of artificial neural networks, and the current desire for reconfigurable flight control are the main motivations for the work described here.

This chapter develops a new adaptive control architecture using evolutionary algorithms, specifically neural networks and genetic algorithms. The control architecture is tested on the adaptive attitude control problem, and simulation results are included for this problem.

8.2 The Adaptive Control Architecture

The rigid body satellite is described by the Euler equations (3.24) and (3.32). The problem faced here is, its detumbling and reorientation in a prespecified way. This needs to be done, even in cases where its dynamics are unknown, due to damage, malfunction or a change in the moments of inertia.

Assume that the dynamics of the satellite are unknown or have changed for an unknown reason (which could be one of the above). An adaptive control architecture known as *neuro-genetic control* (as introduced in [66]), is shown in Figures 8.1 and 8.2. As soon as the satellite dynamics change, a model of its dynamics is constructed. This is done by adapting (training) a neural network model to identify the plant dynamics (Figure 8.1). At each time instant, the difference between the correct plant output, and the network's estimated plant output, is used as a training error for the neuromodel (as described in the block diagram of Figure 8.1). This method of "neuromodeling" different dynamic systems has already been described and tested in Chapter 5.

The controller is based on genetic algorithms. Assuming that the goal of the control application is to detumble the satellite, and spin it about one of the body axes, while bringing it into a desired orientation, the following objective function is defined:

$$U(t) = |\dot{\Theta} + \Theta| + |\Theta| + |\dot{\Phi} + \Phi| + |\Phi| \tag{8.1}$$

The genetic controller uses the constructed plant model as a predictor of the future values of the plant (block diagram of Figure 8.2). According to the above objective function, which has to be minimized, the goal is to spin the satellite about the z body axis, whilst detumbling it about the x, y axes (make ω_1, ω_2 equal to zero), and reorient it to the position $\Theta = 0, \Phi = 0$. It must be noted, that if the goal was a different one, then it would suffice to change the objective function. The genetic controller applying the genetic algorithm described in Section 7.6 minimizes (8.1) by maximizing the following adjusted fitness function:

$$E(t) = \frac{1}{1 + U(t)} \tag{8.2}$$

Thus at each time instant, the genetic controller seeks for individuals (chromosomes) of the form in Figure 7.11, which lead the system is a "good" state, according to the objectives (8.1).

This chapter shows examples where the goal is to reach a desired target state of the plant from another initial state. Generally, this problem can be

8.3. LPN FOR THE ADAPTIVE CONTROL ARCHITECTURE

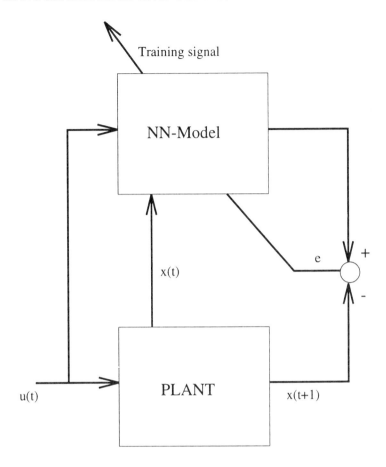

Figure 8.1: Building the neuromodel for the plant. (From *Neural Networks and their Applications, J.G.Taylor [Ed.]*. Copyright John Wiley & Sons Limited. Reproduced with permission.)

considered more difficult than the problem in which there is a reference model to follow (tracking problem). In that case (which is a simpler one), it suffices to change the objective function (8.1) accordingly, by taking its difference with the reference model.

8.3 The Use of LPN for the Adaptive Control Architecture

As shown in Chapter 5, artificial neural networks have the capability of constructing accurate local models of complex dynamic systems over a large region

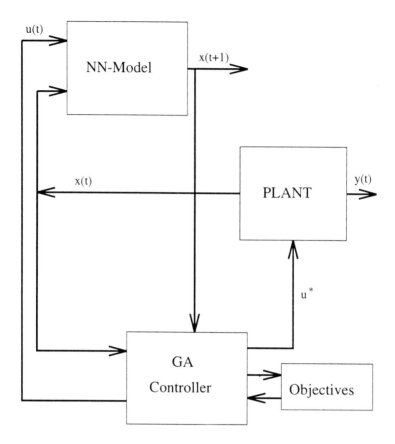

Figure 8.2: The genetic controller. (From *Neural Networks and their Applications*, J.G. Taylor [Ed.]. Copyright John Wiley & Sons Limited. Reproduced with permission.)

of the phase space. This capability of local predictive networks (LPN) is essential for the adaptive control architecture described in the previous Section.

In this section, another example of local prediction using LPN networks is given. The Euler equations (3.24) are considered again. This time, the LPN constructed will have as additional inputs the control thrusts of the satellite.

The values of the dynamic system (3.24) were chosen to be $I_x = 3, I_y = 2, I_z = 1$. A LPN was trained to predict ahead for $\Delta t = 0.5$, using 501 data taken from the trajectory starting at $(\omega_1, \omega_2, \omega_3) = (3, 2, 1)$ with $L = -1.2\omega_1 + \frac{\sqrt{6}}{2}\omega_3$, $M = 0.35\omega_2$, $N = -\sqrt{6}\omega_1 - 0.4\omega_3$. Thus, the training data were chosen from the chaotic trajectory described in Chapters 3 and 5. The network architecture was 24-20-10-3, since a value of $p = 3$ in equation (5.3)

8.3. LPN FOR THE ADAPTIVE CONTROL ARCHITECTURE

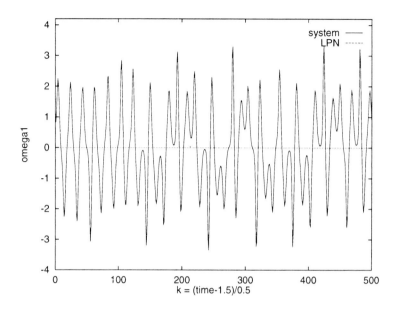

Figure 8.3: Euler equations. How well the 24-20-10-3 LPN learnt the training data - angular velocity about x principal axis. LPN predicting 0.5 sec ahead.

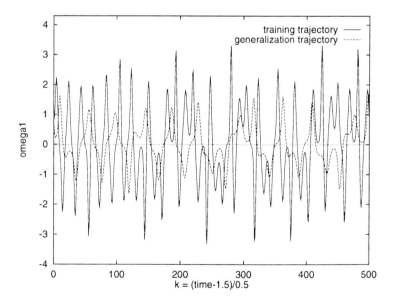

Figure 8.4: Euler equations - angular velocity about x principal axis. System trajectory used for training and system trajectory used for the generalization test of the trained LPN.

was chosen, and the network had 12 inputs describing the current and 3 past states of the system and another 12 inputs describing the current and 3 past control inputs to the system.

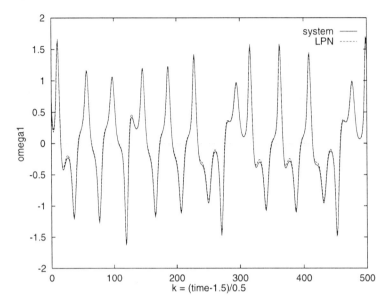

Figure 8.5: Euler equations - angular velocity about x principal axis. How well the 24-20-10-3 LPN generalizes on the trajectory starting at $(3, 2, 1)$ with $L = -1.5\omega_1 + 0.5\omega_3$, $M = 0.2\omega_2$, $N = -2\omega_1 - 0.3\omega_3$. LPN predicting 0.5 sec ahead.

The training convergence of the above LPN (angular velocity ω_1) is shown in Figure 8.3 while the generalization capability of the network was tested in a trajectory started at $(\omega_1, \omega_2, \omega_3) = (3, 2, 1)$ with $L = -1.5\omega_1 + \frac{1}{2}\omega_3$, $M = 0.2\omega_2$, $N = -2\omega_1 - 0.3\omega_3$ which is a trajectory completely different from the training one (see Figure 8.4). The result of the generalization test is shown in Figure 8.5 (angular velocity ω_1). The results confirm those obtained in Chapter 5.

It should be noted that for the neuro-genetic adaptive control architecture described in the previous section and applied throughout this Chapter, a prediction interval of $\Delta t = 0.5$ is not required. Indeed since the time step at which the satellite operations occur (time at which the control thrusts of the satellite can change) is $\Delta t = 0.01$, prediction is required only for 0.01 sec ahead. Thus the prediction interval described above of $\Delta t = 0.5$ is 50 times larger than required, but the accuracy is still good as can be seen from the diagrams.

The results of Chapter 5 and the results in this section illustrate that a

8.4 Spin Stabilization of a Satellite About a Stable Axis

Assume that for some unknown reason (damage), a satellite with specified dynamics changes its characteristics. Its moments of inertia, become $I_x = 1160 Kg \cdot m^2$, $I_y = 23300 Kg \cdot m^2$ and $I_z = 24000 Kg \cdot m^2$. During the period where

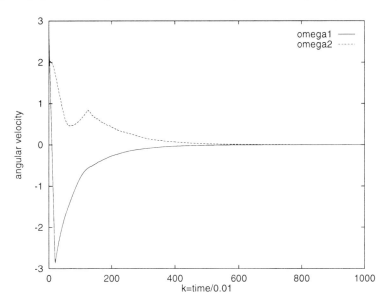

Figure 8.6: Angular velocities ω_1, ω_2 for the satellite after the application of the genetic controller. (From *Neural Networks and their Applications*, J.G. Taylor [Ed.]. Copyright John Wiley & Sons Limited. Reproduced with permission.)

the system changes dynamics, unknown forces lead it to the $(\omega_1, \omega_2, \omega_3) = (3, 2, 1)$ and $(\Theta, \Phi, \Psi) = (1, 2, 3)$ state. The goal is to detumble the satellite about the x, y body axes, spin it about the z body axis, and reorient it, so that $\Theta = 0, \Phi = 0$.

Note: Neural network can be trained to predict locally and accurately over a large part of the phase space, even when the system dynamics are highly nonlinear and chaotic. In the simulation results that follow in this Chapter, every time the system dynamics change in an unknown way, it will be assumed that a neuromodel describing the new dynamics of the plant has been trained according to the adaptive control architecture described in section 8.2.

The application of the genetic adaptive controller architecture described by Figures 8.1, 8.2 leads to the situation described by Figures 8.6-8.11. Figure 8.6 shows the evolution in time of the angular velocities ω_1, ω_2 about the x, y body axes respectively. It is easily shown that the genetic controller soon leads both of these angular velocities to the prespecified value of zero. The third angular velocity is arbitrary, since it was not in the control objectives.

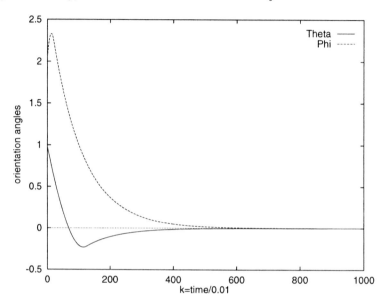

Figure 8.7: Orientation angles Θ, Φ for the satellite after the application of the genetic controller. (From *Neural Networks and their Applications, J.G. Taylor [Ed.]*. Copyright John Wiley & Sons Limited. Reproduced with permission.)

Figure 8.7 show the reorientation of the satellite for the angles Θ, Φ after the application of the controller. While these are becoming zero, the satellite is rotating about its z axis (Figure 8.8). It should be noted that the controller not only leads the system to a desired state, but it maintains this state afterwards. The applied thrusts during the genetic control of the satellite are shown in Figures 8.9, 8.10, and 8.11. It can be noticed, that during the time that the angular velocities are large the thrusts vary very rapidly, so that in some situations they can be seen as applying a kind of bang-bang control. As soon as the angular velocities obtain small values, the required and applied thrusts L, M, become "smooth". The third thrust N is not subject to evolutionary pressure (near the target state) and consequently remains large. This is a result of the particular choice of objective function since the target state

8.4. SATELLITE SPIN STABILIZATION ABOUT A STABLE AXIS   141

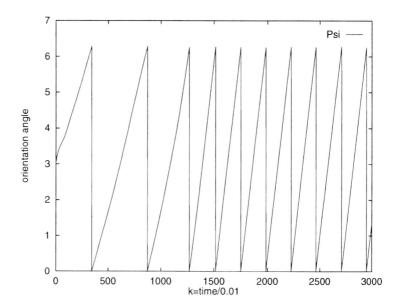

Figure 8.8: Orientation angle Ψ for the satellite after the application of the genetic controller. Initial conditions $(\omega_1, \omega_2, \omega_3) = (3, 2, 1)$ and $(\Theta, \Phi, \Psi) = (1, 2, 3)$.

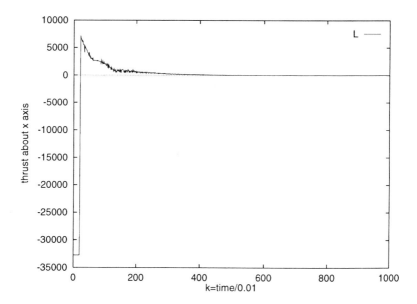

Figure 8.9: Applied thrust L about the x body axis for the satellite during the application of the genetic controller. Initial conditions $(\omega_1, \omega_2, \omega_3) = (3, 2, 1)$ and $(\Theta, \Phi, \Psi) = (1, 2, 3)$.

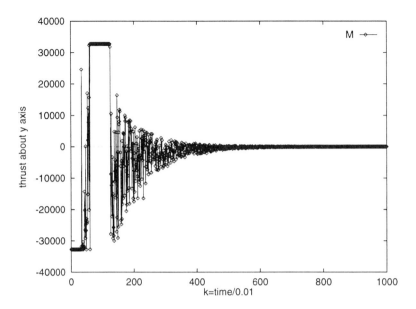

Figure 8.10: Applied thrust M about the y body axis for the satellite during the application of the genetic controller. Initial conditions $(\omega_1, \omega_2, \omega_3) = (3, 2, 1)$ and $(\Theta, \Phi, \Psi) = (1, 2, 3)$.

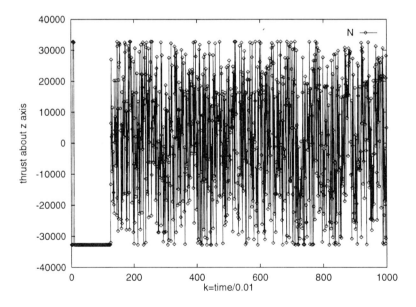

Figure 8.11: Applied thrust N about the z body axis for the satellite during the application of the genetic controller. Initial conditions $(\omega_1, \omega_2, \omega_3) = (3, 2, 1)$ and $(\Theta, \Phi, \Psi) = (1, 2, 3)$.

8.5 Spin Stabilization of a Satellite About an Unstable Axis

In this section the task of stabilizing a satellite (rigid body) about an unstable axis is considered. The moments of inertia of the satellite are $I_x = 1160 Kg \cdot m^2$, $I_y = 24000 Kg \cdot m^2$ and $I_z = 23300 Kg \cdot m^2$. The z body principal axis is *unstable*, since it corresponds to the intermediate principal axis ($I_x < I_z < I_y$) [10, 19]. In this case if the body is rotating about its z axis, a very small disturbance may produce a very great change in the subsequent motion, and the rotation is *unstable* [172, 22].

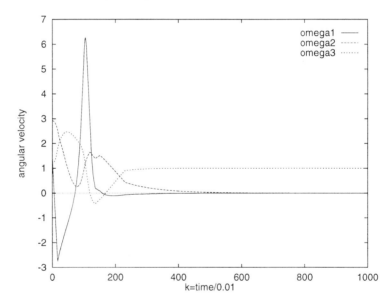

Figure 8.12: Angular velocities $\omega_1, \omega_2, \omega_3$ for the satellite after the application of the genetic controller. Spin stabilization about an unstable axis. Initial conditions for the system are $(\omega_1, \omega_2, \omega_3) = (1.7494, 2.9092, 1.3276)$ and $(\Theta, \Phi, \Psi) = (1, 2, 3)$.

The goal in this simulation is to spin stabilize the satellite about its z (unstable) axis. Thus the target state of the system is set to $(\omega_1, \omega_2, \omega_3) = (0, 0, 1.0)$ and $(\Theta, \Phi, \Psi) = (0, 0, 0)$. According to this a fixed spin of $\omega_3 = 1.0$

is specified. The initial conditions are: $(\omega_1, \omega_2, \omega_3) = (1.7494, 2.9092, 1.3276)$ and $(\Theta, \Phi, \Psi) = (1.0, 2.0, 3.0)$.

The objective function for the genetic controller is specified to be:

$$U(t) = |\dot{\Theta} + \Theta| + |\Theta| + |\dot{\Phi} + \Phi| + |\Phi| + |\dot{\Psi} - 1.0| \quad (8.3)$$

Figure 8.12 shows the evolution in time of the angular velocities about the x, y, z body axes respectively, after the application of the genetic controller. It is easily shown that the genetic controller soon leads the two of them to the prespecified value of zero and the third to the prespecified value of one. As soon as this is achieved, the controller maintains these angular velocities.

In Figure 8.13 the reorientation of the satellite for the angles Θ, Φ after the application of the controller is shown. While these are becoming zero, the

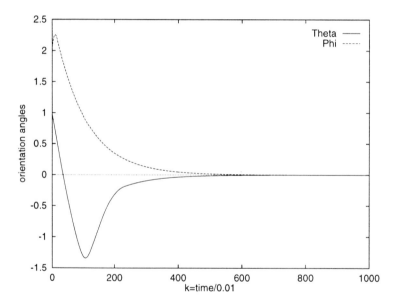

Figure 8.13: Orientation angles Θ, Φ for the satellite after the application of the genetic controller. Spin stabilization about an unstable axis. The initial conditions for the system are $(\omega_1, \omega_2, \omega_3) = (1.7494, 2.9092, 1.3276)$ and $(\Theta, \Phi, \Psi) = (1, 2, 3)$.

satellite is rotating about its z axis (Figure 8.14). It can be noticed that the controller not only reorients the satellite, but it also maintains this reorientation.

8.5. STABILIZATION ABOUT AN UNSTABLE AXIS

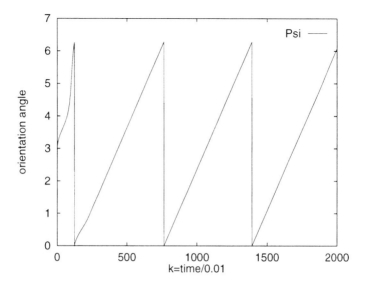

Figure 8.14: Orientation angle Ψ for the satellite after the application of the genetic controller. Spin stabilization about an unstable axis. Initial conditions for the system are $(\omega_1, \omega_2, \omega_3) = (1.7494, 2.9092, 1.3276)$ and $(\Theta, \Phi, \Psi) = (1, 2, 3)$.

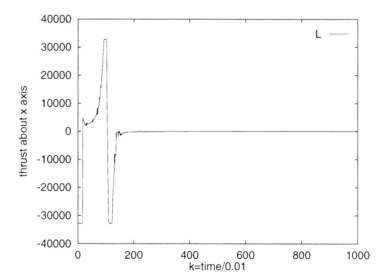

Figure 8.15: Applied thrust L about the x body axis for the satellite during the application of the genetic controller. Spin stabilization about an unstable axis. Initial conditions $(\omega_1, \omega_2, \omega_3) = (1.7494, 2.9092, 1.3276)$ and $(\Theta, \Phi, \Psi) = (1, 2, 3)$.

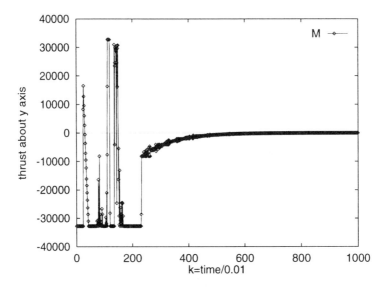

Figure 8.16: Applied thrust M about the y body axis for the satellite during the application of the genetic controller. Spin stabilization about an unstable axis. Initial conditions $(\omega_1, \omega_2, \omega_3) = (1.7494, 2.9092, 1.3276)$ and $(\Theta, \Phi, \Psi) = (1, 2, 3)$.

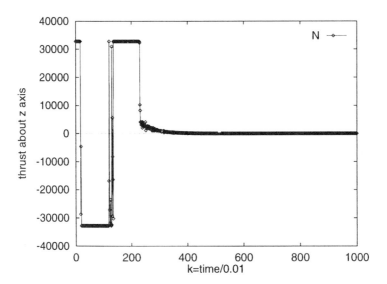

Figure 8.17: Applied thrust N about the z body axis for the satellite during the application of the genetic controller. Spin stabilization about an unstable axis. Initial conditions $(\omega_1, \omega_2, \omega_3) = (1.7494, 2.9092, 1.3276)$ and $(\Theta, \Phi, \Psi) = (1, 2, 3)$.

8.6. STABILIZATION SUBJECT TO SENSOR NOISE   147

The applied thrusts during the genetic control of the satellite are shown in Figures 8.15, 8.16, and 8.17. It should be noticed that the thrusts are not smooth at all.

8.6 Spin Stabilization of a Satellite Subject to Sensor Noise

In this section the robustness of the described adaptive control architecture is tested. For this reason artificial simulated noise is introduced into the sensors of the satellite. The moments of inertia of the satellite are $I_x = 1160 Kg \cdot m^2$, $I_y = 23300 Kg \cdot m^2$ and $I_z = 24000 Kg \cdot m^2$.

The goal in this simulation is to spin stabilize the satellite about its z axis which is stable since it corresponds to the largest moment of inertia. Thus, the target state of the system is set to $(\omega_1, \omega_2, \omega_3) = (0, 0, 1.0)$ and $(\Theta, \Phi, \Psi) = (0, 0, 0)$. The initial conditions are: $(\omega_1, \omega_2, \omega_3) = (1.8, 2.7, 1.4)$ and $(\Theta, \Phi, \Psi) = (1.0, 2.0, 3.0)$. Noise (following the uniform distribution) of 10% of the current sensor values is inserted in the sensors (i.e. all the orientation angles and the

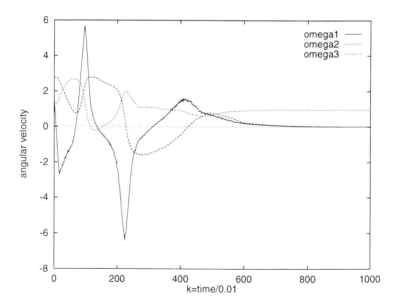

Figure 8.18: Angular velocities $\omega_1, \omega_2, \omega_3$ for the satellite after the application of the genetic controller. Presence of 10% noise in the sensors. The initial conditions for the system are $(\omega_1, \omega_2, \omega_3) = (1.8, 2.7, 1.4)$ and $(\Theta, \Phi, \Psi) = (1, 2, 3)$.

angular velocities about the body fixed axes sensors). Thus, the controller does not have available the real (accurate) state of the plant.

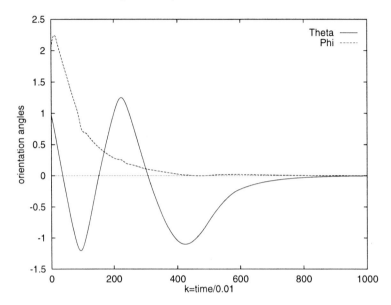

Figure 8.19: Orientation angles Θ, Φ for the satellite after the application of the genetic controller. Presence of 10% noise in the sensors. Initial conditions for the system are $(\omega_1, \omega_2, \omega_3) = (1.8, 2.7, 1.4)$ and $(\Theta, \Phi, \Psi) = (1, 2, 3)$.

In this case where sensor noise is present, from the perspective of the controller, the target state will never be reached precisely. Thus we might expect that if on-off control is not applied the thruster activity might remain at an energetic level despite the fact that only small variations of target states should happen. To overcome this it is necessary to modify the objective function to encourage the use of relatively small thrusts on the z body axis in the face of small variations (since the target has $\omega_3 = 1$ the noise in the coordinates about the z axis has a larger absolute value than in the other axes) as the system approaches the target state. Therefore the following objective function is chosen:

$$U(t) = |\dot{\Theta} + \Theta| + |\Theta| + |\dot{\Phi} + \Phi| + |\Phi| + |\dot{\Psi} - 1.0| + 0.01 \cdot |\ddot{\Psi} + \dot{\Psi} - 1.0| \tag{8.4}$$

The coefficient 0.01 in front of the last term helps to speed up the convergence of the controller especially in the beginning of the control. This happens because angular acceleration considerations should play an important rôle only

8.6. STABILIZATION SUBJECT TO SENSOR NOISE

as the system state is sufficiently close to the target state. Other simulations showed that larger values for the above coefficient tend to slow down the control convergence in the beginning, while later on, when the plant state is close to the target state they help to control the plant using less energy.

Figure 8.18 shows the evolution in time of the angular velocities about the x, y, z body axes respectively, after the application of the genetic controller. Despite the significant amount of noise, the genetic adaptive controller leads the two of them to the prespecified value of zero and the third to the prespecified value of one quite accurately.

In Figure 8.19 the reorientation of the satellite for the angles Θ, Φ after the application of the controller is shown. While these are becoming zero, the satellite is rotating about its z axis (Figure 8.20).

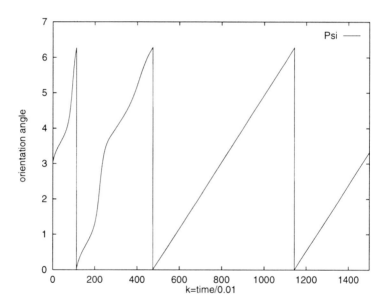

Figure 8.20: Orientation angle Ψ for the satellite after the application of the genetic controller. Presence of 10% noise in the sensors. Initial conditions for the system are $(\omega_1, \omega_2, \omega_3) = (1.8, 2.7, 1.4)$ and $(\Theta, \Phi, \Psi) = (1, 2, 3)$.

The applied thrusts during the genetic control of the satellite are shown in Figures 8.21, 8.22, and 8.23. All of the diagrams in this section confirm that the genetic controller is robust to a significant amount of noise.

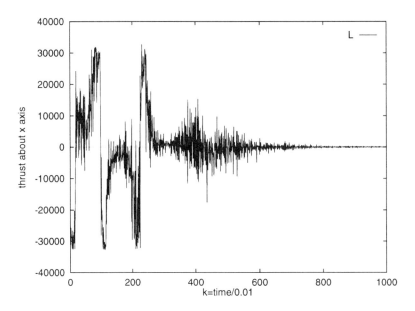

Figure 8.21: Applied thrust L about the x body axis for the satellite during the application of the genetic controller. Presence of 10% noise in the sensors. Initial conditions $(\omega_1, \omega_2, \omega_3) = (1.8, 2.7, 1.4)$ and $(\Theta, \Phi, \Psi) = (1, 2, 3)$.

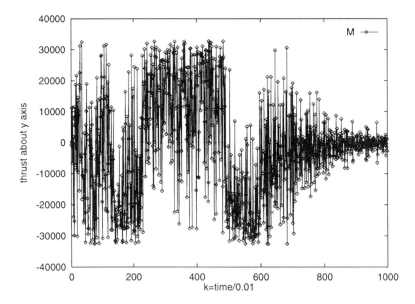

Figure 8.22: Applied thrust M about the y body axis for the satellite during the application of the genetic controller. Presence of 10% noise in the sensors. Initial conditions $(\omega_1, \omega_2, \omega_3) = (1.8, 2.7, 1.4)$ and $(\Theta, \Phi, \Psi) = (1, 2, 3)$.

8.7. ATTITUDE CONTROL SUBJECT TO CHAOS

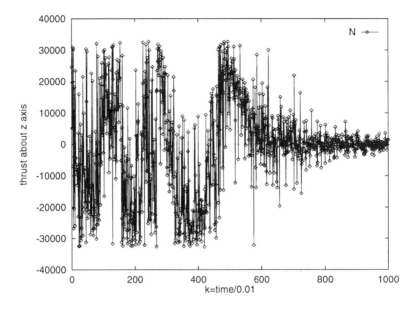

Figure 8.23: Applied thrust N about the z body axis for the satellite during the application of the genetic controller. Presence of 10% noise in the sensors. Initial conditions $(\omega_1, \omega_2, \omega_3) = (1.8, 2.7, 1.4)$ and $(\Theta, \Phi, \Psi) = (1, 2, 3)$.

8.7 Satellite Attitude Control Subject to External Forces Leading to Chaos

Assume that a satellite changes its characteristic moments of inertia to $I_x = 1160 Kg \cdot m^2, I_y = 23300 Kg \cdot m^2$ and $I_z = 24000 Kg \cdot m^2$. External forces are asked upon it for $t > 0$. The system is described by dynamic equations (3.32) and

$$\begin{aligned} I_x \dot{\omega}_1 - (I_y - I_z)\omega_2\omega_3 &= F_1 + L \\ I_y \dot{\omega}_2 - (I_z - I_x)\omega_3\omega_1 &= F_2 + M \\ I_z \dot{\omega}_3 - (I_x - I_y)\omega_1\omega_2 &= F_3 + N \end{aligned} \quad (8.5)$$

where the torques F_1, F_2, F_3 produced by the external forces are given by the equations

$$\begin{aligned} F_1 &= -1200\omega_1 + 1000 \cdot \frac{\sqrt{6}}{2}\omega_3 \\ F_2 &= 350\omega_2 \\ F_3 &= -1000 \cdot \sqrt{6}\omega_1 - 400\omega_3 \end{aligned} \quad (8.6)$$

The above is a system in which the externally imposed torques F_1, F_2, F_3 would, if left to themselves, result in a chaotic motion (see section 3.2), while the thrust vector $G = (L, M, N)$ is trying to control this chaotic motion and lead the system into a prespecified state. In this case, it is assumed that the initial state is $(\omega_1, \omega_2, \omega_3) = (1.3, 3.0, 2.8)$ and $(\Theta, \Phi, \Psi) = (1, 2, 3)$ while the desired target state is $(\omega_1, \omega_2, \omega_3) = (0, 0, 0)$ and $(\Theta, \Phi, \Psi) = (0, 0, 0)$. Since the goal from the previous simulations is different, a new objective function is defined as follows:

$$U(t) = |\dot{\Theta} + \Theta| + |\Theta| + |\dot{\Phi} + \Phi| + |\Phi| + |\dot{\Psi} + \Psi| + |\Psi| \qquad (8.7)$$

Figure 8.24, show the evolution in time of the angular velocities $\omega_1, \omega_2, \omega_3$ about the x, y, z body axes respectively. The genetic controller soon leads these angular velocities to the prespecified value of zero despite the fact that during the control process the external perturbing torques were as large as 44% of the maximum available control torques. As soon as this is achieved, the controller maintains these angular velocities.

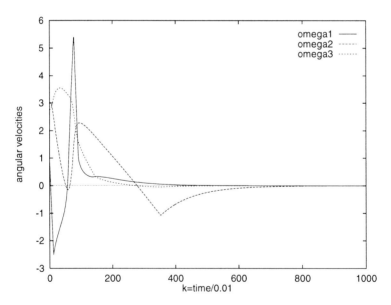

Figure 8.24: Angular velocities ω_1, ω_2, ω_3 for the satellite, in the presence of forces trying to lead it in a chaotic motion, after the application of the genetic controller. Initial conditions $(\omega_1, \omega_2, \omega_3) = (1.3, 3.0, 2.8)$ and $(\Theta, \Phi, \Psi) = (1, 2, 3)$. (From *Neural Networks and their Applications*, J.G.Taylor [Ed.]. Copyright John Wiley & Sons Limited. Reproduced with permission.)

8.7. ATTITUDE CONTROL SUBJECT TO CHAOS

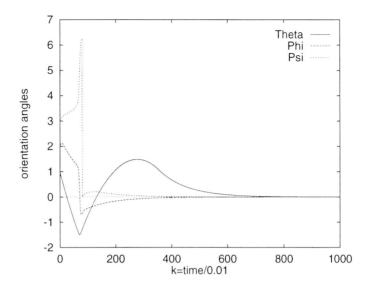

Figure 8.25: Orientation angles Θ, Φ, Ψ for the satellite, in the presence of forces trying to lead it in a chaotic motion, after the application of the genetic controller. Initial conditions $(\omega_1, \omega_2, \omega_3) = (1.3, 3.0, 2.8)$ and $(\Theta, \Phi, \Psi) = (1, 2, 3)$. (From *Neural Networks and their Applications, J.G. Taylor [Ed.]*. Copyright John Wiley & Sons Limited. Reproduced with permission.)

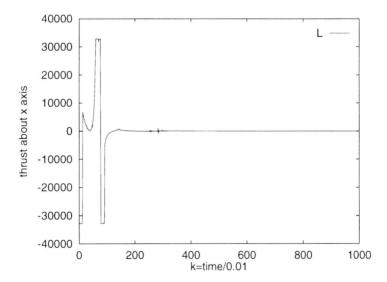

Figure 8.26: Applied thrust L about the x body axis for the satellite, in the presence of forces trying to lead it in a chaotic motion, during the application of the genetic controller. Initial conditions $(\omega_1, \omega_2, \omega_3) = (1.3, 3.0, 2.8)$ and $(\Theta, \Phi, \Psi) = (1, 2, 3)$.

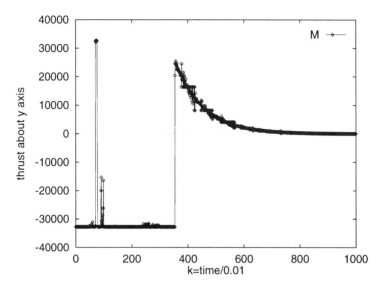

Figure 8.27: Applied thrust M about the y body axis for the satellite, in the presence of forces trying to lead it in a chaotic motion, during the application of the genetic controller. Initial conditions $(\omega_1, \omega_2, \omega_3) = (1.3, 3.0, 2.8)$ and $(\Theta, \Phi, \Psi) = (1, 2, 3)$.

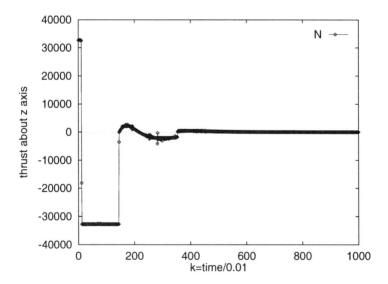

Figure 8.28: Applied thrust N about the z body axis for the satellite, in the presence of forces trying to lead it in a chaotic motion, during the application of the genetic controller. Initial conditions $(\omega_1, \omega_2, \omega_3) = (1.3, 3.0, 2.8)$ and $(\Theta, \Phi, \Psi) = (1, 2, 3)$.

In Figure 8.25 the reorientation of the satellite for the angles Θ, Φ, Ψ after the application of the controller is shown. Again it can be noticed that the controller not only reorients the satellite, but it maintains this reorientation.

The applied thrusts during the genetic control of the satellite are shown in Figures 8.26, 8.27, and 8.28.

8.8 Control Subject to Sensor Noise and External Forces Leading to Chaos

This section is another test for the robustness of the adaptive control architecture described in section 8.2. Artificial noise following the uniform distribution is introduced to the sensors to the amount of 5% of the current sensor values (i.e. all the orientation angles and the angular velocities about the three principal axes' sensors).

As in the previous simulation experiment, external forces are acting upon the satellite for $t > 0$. The system is again described by dynamic equations (3.32), (8.5) and (8.6). The plant is a system in which the externally imposed torques F_1, F_2, F_3 would, if left to themselves, result in a chaotic motion (see section 3.2), while the thrust vector $G = (L, M, N)$ is trying to control this chaotic motion and lead the system into a prespecified state. In this case, the satellite moments of inertia are $I_x = 1200 Kg \cdot m^2, I_y = 22000 Kg \cdot m^2, I_z = 25000 Kg \cdot m^2$ (something which is not known to the controller) and the initial state is $(\omega_1, \omega_2, \omega_3) = (1.5, 2.5, 1.2)$ and $(\Theta, \Phi, \Psi) = (1, 2, 3)$ while the desired target state is $(\omega_1, \omega_2, \omega_3) = (0, 0, 1.0)$ and $(\Theta, \Phi, \Psi) = (0, 0, 0)$. Since noise is present, in order to reduce the energy spend by the controller in the later stages of control (the plant is close to the target state, so much of the error is due to noise) the objective function adopted in section 8.6 is used:

$$U(t) = |\dot{\Theta} + \Theta| + |\Theta| + |\dot{\Phi} + \Phi| + |\Phi| + |\dot{\Psi} - 1.0| + 0.01 \cdot |\ddot{\Psi} + \dot{\Psi} - 1.0| \tag{8.8}$$

As previously (section 8.6), the coefficient 0.01 in front of the last term in the objective function speeds up the control procedure since the angular acceleration term is not considered in the early stages of the control.

Figure 8.29 show the evolution in time of the angular velocities $\omega_1, \omega_2, \omega_3$ about the x, y, z body axes respectively. Although the controller does not have available the accurate values of the current plant state (since there is a

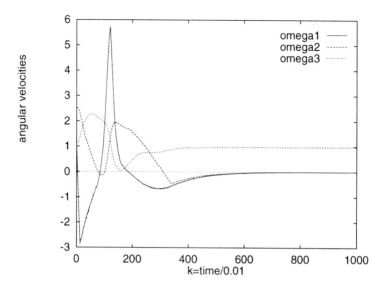

Figure 8.29: Angular velocities ω_1, ω_2, ω_3 for the satellite, in the presence of forces trying to lead it in a chaotic motion, after the application of the genetic controller. Presence of 5% noise in the sensors. Initial conditions $(\omega_1, \omega_2, \omega_3) = (1.5, 2.5, 1.2)$ and $(\Theta, \Phi, \Psi) = (1, 2, 3)$.

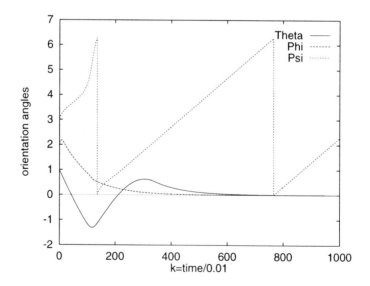

Figure 8.30: Orientation angles Θ, Φ, Ψ for the satellite, in the presence of forces trying to lead it in a chaotic motion, after the application of the genetic controller. Presence of 5% noise in the sensors. Initial conditions $(\omega_1, \omega_2, \omega_3) = (1.5, 2.5, 1.2)$ and $(\Theta, \Phi, \Psi) = (1, 2, 3)$.

8.8. ATTITUDE CONTROL SUBJECT TO CHAOS AND NOISE

significant amount of noise), it manages to lead ω_1, ω_2 to the prespecified value of zero, and ω_3 to the target value of one despite the presence of large perturbing forces. As soon as this is achieved with reasonable accuracy (accuracy above a certain point cannot be achieved due to noise), the controller maintains these angular velocities.

In Figure 8.30 the reorientation of the satellite for the angles Θ, Φ, Ψ after the application of the controller is shown. Again it can be noticed that the controller not only reorients the satellite to the target orientation $(0,0,0)$ with acceptable accuracy (since there is 10% noise accuracy beyond a level cannot be achieved), but it maintains this reorientation.

The applied thrusts during the genetic control of the satellite are shown in Figures 8.31, 8.32, and 8.33. All the diagrams show the control thrusts in the time interval [0,10].

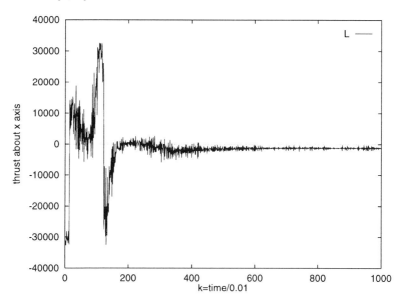

Figure 8.31: Applied thrust L about the x body axis for the satellite, in the presence of forces trying to lead it in a chaotic motion, during the application of the genetic controller. Presence of 5% noise in the sensors. Initial conditions $(\omega_1, \omega_2, \omega_3) = (1.5, 2.5, 1.2)$ and $(\Theta, \Phi, \Psi) = (1, 2, 3)$.

All of the above results confirm that the adaptive control architecture described in section 8.2 is able to "control chaos" even in the presence of noise. The system seems to be sufficiently robust, a characteristic of intelligent control.

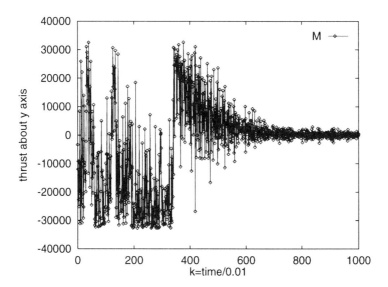

Figure 8.32: Applied thrust M about the y body axis for the satellite, in the presence of forces trying to lead it in a chaotic motion, during the application of the genetic controller. Presence of 5% noise in the sensors. Initial conditions $(\omega_1, \omega_2, \omega_3) = (1.5, 2.5, 1.2)$ and $(\Theta, \Phi, \Psi) = (1, 2, 3)$.

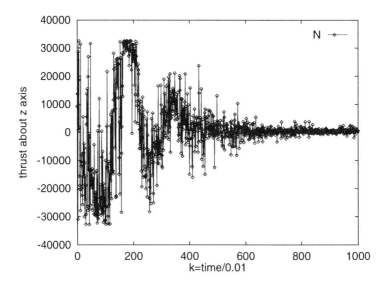

Figure 8.33: Applied thrust N about the z body axis for the satellite, in the presence of forces trying to lead it in a chaotic motion, during the application of the genetic controller. Presence of 5% noise in the sensors. Initial conditions $(\omega_1, \omega_2, \omega_3) = (1.5, 2.5, 1.2)$ and $(\Theta, \Phi, \Psi) = (1, 2, 3)$.

8.9 An Extension to the Adaptive Control Architecture

An extension of the adaptive control architecture described in section 8.2 is shown in Figures 8.34, 8.35 and 8.36. This architecture takes advantage of the fact that hardware implementation of neural networks is possible and, thus, the neurocontroller shown can be faster in operation than the genetic controller. In theory, the two methods are equivalent.

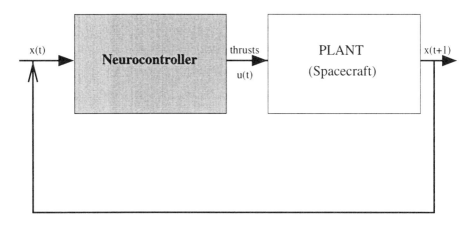

Figure 8.34: Neurocontroller in operation.

Again, as soon as the spacecraft dynamics change characteristics, we construct a model of it, following Figure 8.1. The next step is to construct an optimization module (called the *Instructor*) which is implemented with a genetic algorithm. The goal of the *Instructor* is to build a training set for the neurocontroller (Figure 8.35) by defining the objectives (equation (8.1), for example, if our goal is the same as in section 8.2). Now, the neurocontroller can be trained using the procedure in Figure 8.36 and put into operation (Figure 8.34). Although this neurocontroller requires some time to be trained before becoming operational, the assumption of hardware implementation of the neural network makes this time small. Even when this assumption is not entirely valid, the adaptive neurocontroller architecture can be applied by considering a simple controller to control the spacecraft roughly, as long as the neurocontroller is not trained. In this case, it is assumed that while the neurocontroller is being trained, the plant is not led to destruction.

It is worth noting, that the above procedure for constructing the train-

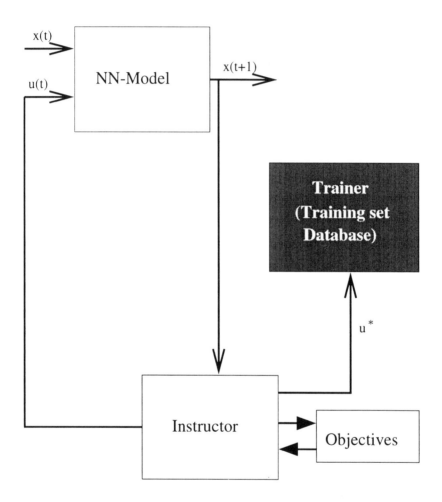

Figure 8.35: Building the trainer of the neurocontroller.

8.10. DISCUSSION

ing set of the neurocontroller solves the ill-posedness problem of some plants, something which is a principal source of difficulty in applying some other neurocontrol techniques, as discussed in Chapter 4.

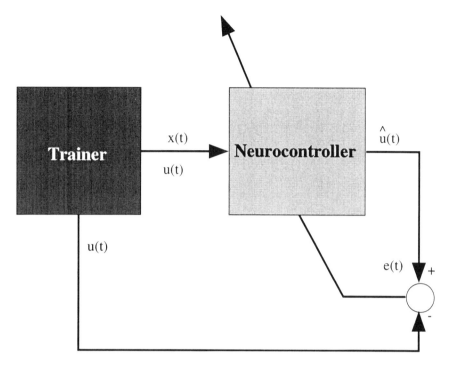

Figure 8.36: Training the adaptive Neurocontroller.

8.10 Discussion

The adaptive control architecture, described in this chapter, is able to tackle the attitude control of a rigid body. The method deals directly with the underlying nonlinearities describing the motion of the rigid body, rather than with linearized equations (an approach usually adopted by other attitude control methods), which can lead to inaccurate or incorrect description of the system dynamics.

In contrast to most other attitude control approaches, it deals with the adaptive problem. In cases where the system dynamics change, due to damage or malfunction, the control architecture described here adapts itself to the new plant dynamics, and where possible it tries to control the new system by constructing a model of the plant dynamics.

In many real applications, there is no *a priori* reference model which must be followed; instead, it is important to choose current actions so as to optimize an end result at a later time [215]. Thus, the conventional model reference adaptive control methods (MRAC) are not generally applicable to some control problems (for example, if the goal is to achieve a special end state), due to the possible lack of a reference model. The described adaptive control architecture, can be considered general, and applicable to many control problems. It does not require a reference model, however, when the problem is to follow a specified reference model, the neuro-genetic method can be applied by changing the objective function.

The simulations showed that the overall control system is robust to a significant amount of noise and stable, at least for the cases where the goal is to spin the spacecraft about a body axis, while bringing it to a desired orientation, or completely detumble the spacecraft (all angular velocities are zero) while reorienting it in a pre-specified position according to an inertial axes frame. In addition, it was shown that it is possible to control a system, even in the case where the motion is chaotic, under the application of forces which lead the system to chaos. This control of chaos can be very advantageous and most important in many applications.

Compared with the OGY method described in Chapter 2 the control architecture presented here can effectively control a chaotic system starting from any system state. In contrast, the OGY method has to wait until the system trajectory passes through an unstable periodic orbit in order to start control [55, 156]. In addition, the OGY method requires some knowledge of the plant dynamics (i.e. unstable periodic orbits) while the neuro-generic method requires no minimum knowledge of the plant dynamics. Unlike the OGY method which belongs to the category of 'weak methods', the neuro-genetic control architecture is a 'strong method' which aims to employ comparatively large variations of control signals to force the system to acquire any desired target state or to track any desired trajectory in state space.

Compared with other current neurocontrol techniques, the method is able to find a particular solution of the inverse kinematics problem, even in cases where the problem is ill-posed. It is goal-directed (all actions are according and towards a specified goal), and it can be applied for either learning an inverse model, or trajectory tracking, or reaching a particular state of the plant. However, it should be noted that its application to various problems such as the above, requires the correct choice of the objective function. An inappropriate

8.10. DISCUSSION

choice of the objective function can lead to bad or unstable results. It is not always easy or obvious how to choose the objective function.

Chapter 9

Conclusions and the Future

Every solution breeds new problems
Anonymous

This work presented some techniques for nonlinear adaptive control using neural networks and genetic algorithms.

In the course of developing adaptive control methods, using plant models, techniques for modeling dynamic systems of varying complexity using neural networks were also investigated. Supervised learning neural networks perform well in this task of modeling. The constructed models are able to predict accurately for a short time ahead. By training these neuromodels on a single trajectory, good local prediction can be achieved over a large part of the dynamic system phase space.

A new adaptive control architecture was fully described using the attitude control problem as a testbed. This nonlinear problem is complicated, and as shown becomes more complicated when the dynamics of the system become chaotic under certain conditions.

The implemented adaptive control architecture is general in nature, in the sense that it is applicable to different types of control tasks, such as reaching a particular state of a plant, or tracking an available reference model. Adaptive nonlinear control is achieved by constructing a plant model using supervised learning neural networks. The generality of the method extends its applicability to different control problems, as it does not require simplifying assumptions for the plant, or any linearization of the plant dynamics. Instead, it deals directly with the unknown nonlinear equations of the dynamic system under consideration. This system is able to "predict", since in order to control a plant adaptively, the controller determines the next state of the system for specific inputs. It can also be considered as a general engineering tool, since it can be used to solve the inverse kinematics problem for redundant systems (i.e. the dynamic system has no unique transform in the reverse direction).

The empirical study of the described method for the attitude control problem, using computer simulations, showed that the controller is stable as long as the objective function for the controller is chosen carefully. This architecture is able to detumble a spacecraft, spin it in a desired way, and at the same time reorient it, in a pre-specified target orientation even in the presence of a significant amount of noise in the plant sensors. It is also shown, using the theoretical analysis of Lyapunov functions, that a control solution found by genetic programming for the detumbling part is stable.

An exceptional case of difficulty in the control of the nonlinear system, was introduced by adding external forces acting upon the system, and leading it into a chaotic motion. Even in this case, the control architecture behaved extremely well. The chaotic motion of the body was controlled. The spacecraft was completely detumbled, it was reoriented in a desired orientation, and then it was maintained in that target position.

The attitude control problem which was used as a testbed for the control architecture described in Chapter 8 has not previously been treated in the adaptive case by other attitude control approaches. By employing machine learning techniques, the attitude control problem can be tackled successfully.

Generally, this work highlights some of the possibilities for nonlinear and adaptive control using machine learning techniques, a research area which looks very promising and which is developing very rapidly. There are still many things to be done before establishing a uniformly acceptable framework.

Theoretical stability results should be one of the topics of future research. It is not always easy, or possible to prove *a posteriori* the stability of a control algorithm. Perhaps in the future, an "automatic" method for choosing the objective function of the described control architecture will be developed, so as the overall system serves as a kind of a Lyapunov function. In this way, stability may be guaranteed from theory.

Another field towards which research should be directed, in the next few years, is that of supervised learning algorithms. Current backpropagation algorithms are slow. Their speed of learning, when not implemented in hardware, makes them inapplicable for on-line control. Although better algorithms have already appeared, even these new backpropagation variations are slow for the requirements of software plant control. In addition to the speed problem, learning can also be difficult with the current backpropagation algorithms when the data are not smooth enough. Research towards better learning algorithms should consider this difficulty as well. It is necessary to derive new supervised learning algorithms for non smooth functions, which overcome the "generalization vs speed dilemma".

Brain-like control approaches, able to provide plausible models of the brain, together with good engineering functionality [217] is another route for future research. The brain is the best controller known so far, and its characteristics are more than desirable in control systems. The construction of such an advanced artificial system will also give a useful feedback to neuroscience which aims to explain how the brain works [199].

The field of intelligent control is still very new. For the development of new nonlinear and intelligent adaptive control architectures, techniques from conventional control, machine learning, optimal control and optimization should be combined. The ultimate goal is to have intelligent control systems which control accurately, and adapt on-line very rapidly, to possible unknown changes of the systems dynamics or the environment.

Appendix A

Euler Equations Solutions

Δός μοι πᾶ στῶ καί τάν γᾶν κινήσω
[Give me a place to stand and I will move the earth]

Archimedes
(on the discovery of levers)

A.1 An Analytical Solution for Euler Equations

A.1.1 Special Case I

An analytical solution of the Euler equations (3.24) for a specific case is presented here[1]. Assuming, in a specific problem, $I_y = I_z = I$, $L = 0$, and M, N are constants then the Euler equations take the form

$$I_x \dot{\omega}_1 = 0 \tag{A.1}$$

$$I_y \dot{\omega}_2 - (I - I_x)\omega_3 \omega_1 = M \tag{A.2}$$

$$I_z \dot{\omega}_3 - (I_y - I)\omega_2 \omega_3 = N \tag{A.3}$$

This is a *nonhomogeneous* system of *first order* differential equations and can be solved analytically. From (A.1) we find that

$$I_1 \dot{\omega}_1 = 0 \Rightarrow \omega_1 = c \tag{A.4}$$

where c is a constant.

---
[1] An analytical solution of Euler equations in the case of non presence of torques in terms of elliptic functions can be found in [195].

So now, the following system must be solved

$$\dot{\omega}_2 = \frac{(I - I_x)c}{I}\omega_3 + \frac{M}{I} \tag{A.5}$$

$$\dot{\omega}_3 = \frac{(I_x - I)c}{I}\omega_2 + \frac{N}{I} \tag{A.6}$$

This system written in the general form of a nonhomogeneous system gives

$$\mathbf{x}' = \mathbf{P}(t)\mathbf{x} + \mathbf{g}(t) \tag{A.7}$$

where

$$\mathbf{P}(t) = \begin{pmatrix} 0 & \frac{(I-I_x)c}{I} \\ \frac{(I_x-I)c}{I} & 0 \end{pmatrix} \tag{A.8}$$

and

$$\mathbf{g}(t) = \begin{pmatrix} \frac{M}{I} \\ \frac{N}{I} \end{pmatrix} \tag{A.9}$$

The general solution of system (A.7) can be expressed as

$$\mathbf{x} = \mathbf{x}^{(c)}(t) + \mathbf{x}^{(p)}(t) \tag{A.10}$$

where $\mathbf{x}^{(c)}(t)$ is the general solution of the homogeneous system $\mathbf{x}' = \mathbf{P}(t)\mathbf{x}$, and $\mathbf{x}^{(p)}(t)$ is a particular solution of the nonhomogeneous system (A.7).

The theory of systems of differential equations can be further illuminated by introducing the idea of a fundamental matrix.

Suppose that $\mathbf{x}^{(1)}, \ldots, \mathbf{x}^{(n)}$ form a fundamental set of solutions for the equation

$$\mathbf{x}' = \mathbf{P}(t)x \tag{A.11}$$

on some interval $\alpha < t < \beta$. Then the matrix

$$\Psi(t) = \begin{pmatrix} x_1^{(1)}(t) & \cdots & x_1^{(n)}(t) \\ \vdots & \ddots & \vdots \\ x_n^{(1)}(t) & \cdots & x_n^{(n)}(t) \end{pmatrix}, \tag{A.12}$$

whose columns are the vectors $\mathbf{x}^{(1)}, \ldots, \mathbf{x}^{(n)}$ is said to be a *fundamental matrix* for the system (A.11). Note that any fundamental matrix is nonsingular, since its columns are linearly independent vectors.

A.1. AN ANALYTICAL SOLUTION FOR EULER EQUATIONS

The general solution of equation (A.11) is

$$\mathbf{x} = \Psi(t)\mathbf{c} \tag{A.13}$$

where $\mathbf{c}$ is a constant vector with arbitrary components $c_1, c_2, \ldots, c_n$.

Now the eigenvectors for the matrix $\mathbf{P}(t)$ (equation (A.8)) must be found. Noting that

$$\alpha = \frac{(I - I_x)c}{I} \tag{A.14}$$

we have

$$\begin{vmatrix} -r & \frac{(I-I_x)c}{I} \\ \frac{(I_x-I)c}{I} & -r \end{vmatrix} = r^2 + \alpha^2 = 0 \Rightarrow \left\{ \begin{array}{l} r_1 = \alpha i \\ r_2 = -\alpha i \end{array} \right\} \tag{A.15}$$

Therefore, the corresponding eigenvectors are

$$\xi^{(1)} = \begin{pmatrix} 1 \\ i \end{pmatrix}, \xi^{(2)} = \begin{pmatrix} 1 \\ -i \end{pmatrix}. \tag{A.16}$$

A solution then of the homogeneous system (A.11) is

$$\mathbf{x}^{(1)}(t) = \begin{pmatrix} 1 \\ i \end{pmatrix} e^{\alpha i t} \tag{A.17}$$

$$= \begin{pmatrix} 1 \\ i \end{pmatrix} (\cos \alpha t + i \sin \alpha t) \tag{A.18}$$

$$= \begin{pmatrix} \cos \alpha t \\ -\sin \alpha t \end{pmatrix} + i \begin{pmatrix} \sin \alpha t \\ \cos \alpha t \end{pmatrix}. \tag{A.19}$$

If $\mathbf{x}^{(1)}(t) = \mathbf{w}(t) + i\mathbf{v}(t)$, then the vectors

$$\mathbf{w}(t) = \begin{pmatrix} \cos \alpha t \\ -\sin \alpha t \end{pmatrix}, \tag{A.20}$$

$$\mathbf{v}(t) = \begin{pmatrix} \sin \alpha t \\ \cos \alpha t \end{pmatrix} \tag{A.21}$$

are real value solutions of equation (A.11). It is easy to show that $\mathbf{w}(t)$ and $\mathbf{v}(t)$ are linearly independent solutions.

Then the general solution of the homogeneous system (A.11) is given by

$$\mathbf{x}^{(c)} = c_1 \mathbf{w}(t) + c_2 \mathbf{v}(t), \tag{A.22}$$

where $\mathbf{w}(t)$ and $\mathbf{v}(t)$ are given from (A.20),(A.21).

Now the fundamental matrix for this system is

$$\Psi(t) = \begin{pmatrix} \cos \alpha t & \sin \alpha t \\ -\sin \alpha t & \cos \alpha t \end{pmatrix} \quad (A.23)$$

and the general solution for system (A.7) (see ([25]) is

$$\mathbf{x} = \Psi(t)\mathbf{u}(t) \quad (A.24)$$

where $\mathbf{u}(t)$ satisfies $\Psi(t)\dot{\mathbf{u}}(t) = \mathbf{g}(t)$ or from (A.23),(A.9)

$$\begin{pmatrix} \cos \alpha t & \sin \alpha t \\ -\sin \alpha t & \cos \alpha t \end{pmatrix} \begin{pmatrix} \dot{u}_1 \\ \dot{u}_2 \end{pmatrix} = \begin{pmatrix} \frac{M}{I} \\ \frac{N}{I} \end{pmatrix} \quad (A.25)$$

From system (A.25) we find that

$$\dot{u}_1 = -\frac{N}{I}\sin \alpha t + \frac{M}{I}\cos \alpha t \quad (A.26)$$

$$\dot{u}_2 = \frac{\frac{M}{I} - \dot{u}_1 \cos \alpha t}{\sin \alpha t} \quad (A.27)$$

So now we have

$$u_1 = -\frac{N}{I}\int \sin \alpha t\, dt + \frac{M}{I}\int \cos \alpha t\, dt$$

$$= \frac{N}{I\alpha}\cos \alpha t + \frac{M}{I\alpha}\sin \alpha t + c_1 \quad (A.28)$$

where c_1 is a constant.

$$u_2 = \int \frac{M}{I \sin \alpha t} dt + \int \frac{N}{I}\cos \alpha t\, dt -$$

$$\int \frac{M \cos^2 \alpha t}{I \sin \alpha t} dt$$

$$= \frac{M}{I}\int \frac{1}{\sin \alpha t} dt + \frac{N}{I\alpha}\sin \alpha t -$$

$$\frac{M}{I}\int \frac{\cos^2 \alpha t}{\sin \alpha t} dt$$

Substituting $\tan \frac{\alpha t}{2} = u$ in the first integral and $\tan \frac{\alpha t}{2} = w$ in the second

A.1. AN ANALYTICAL SOLUTION FOR EULER EQUATIONS

we take

$$\begin{aligned}
u_2 &= \frac{2M}{Ia} \int \frac{1+u^2}{(1+u^2)(1-u^2)} du + \frac{N}{Ia} \sin \alpha t - \\
&\quad \frac{M}{I} \int \left(\frac{1-w^2}{1+w^2}\right)^2 \frac{1+w^2}{2w} \frac{2}{a(1+w^2)} dw \\
&= \frac{2M}{Ia} \int \frac{1}{1-u^2} du + \frac{N}{Ia} \sin \alpha t - \\
&\quad \frac{M}{Ia} \int \left(-\frac{4w}{(1+w^2)^2} + \frac{1}{w}\right) dw \\
&= \frac{2M}{Ia} \int \frac{1}{(1+u)(1-u)} du + \frac{N}{Ia} \sin \alpha t - \\
&\quad \frac{M}{Ia} \int \frac{(1-w^2)(1-w^2)}{(1+w^2)(1+w^2)w} dw \\
&= \frac{M}{Ia} \int \left(\frac{1}{1-u} + \frac{1}{1+u}\right) du + \frac{N}{Ia} \sin \alpha t - \\
&\quad \frac{M}{Ia} \int \left(-\frac{4w}{(1+w^2)^2} + \frac{1}{w}\right) dw \\
&= \frac{M}{Ia} (\log(1+u) - \log(1-u)) + \frac{N}{Ia} \sin \alpha t - \\
&\quad \frac{M}{Ia} \left(\frac{2}{1+w^2} + \log w\right) \quad\quad\quad (A.29)
\end{aligned}$$

Substituting the values of u, v in the above equation gives

$$u_2 = \frac{N}{Ia} \sin \alpha t - \frac{M}{Ia} \frac{2}{1+\tan^2 \frac{\alpha t}{2}} + c_2 \quad\quad (A.30)$$

where c_2 is a constant.

Thus, from (A.24) the general solution for system (A.5),(A.6) is obtained:

$$\begin{aligned}
\omega_1 &= c \\
\omega_2 &= -u_1 \sin \alpha t + u_2 \cos \alpha t \quad\quad (A.31) \\
\omega_3 &= u_1 \cos \alpha t + u_2 \sin \alpha t
\end{aligned}$$

where u_1 and u_2 are given from equations (A.28),(A.30).

A.1.2 Special Case II

The Euler equations are solved again, with the same data as the previous case, but now with the assumption that M, N are not constant but functions of t. Specifically the Euler equations with $M = \sin \alpha t, N = \cos \alpha t$ are solved.

Using the same method applied in the special case I the following values are obtained:

$$\begin{aligned}
\omega_1 &= c \\
\omega_2 &= c_1 \cos \alpha t + \frac{t \sin \alpha t}{I} + c_2 \sin \alpha t \\
\omega_3 &= \frac{t \cos \alpha t}{I} - c_1 \sin \alpha t + c_2 \cos \alpha t
\end{aligned} \qquad (A.32)$$

where c, c_1, c_2 are constants and α is given by (A.14).

This special set of solutions (A.28),(A.30) and (A.32) for the Euler equations (3.24) will be used to test the accuracy of the simulator. The simulator uses numerical methods (Adams - Moulton predictor-corrector), so it is necessary to compare its performance with an analytical solution of the system, for a specific set of data.

A.2 Computing a Solution for Euler Equations

A simulator for a model describing the motion of a rigid body was designed and used for the experiments of this book.

This simulator takes as input an initial state of the model $\omega^{(t_0)} = (\omega_1^{(t_0)}, \omega_2^{(t_0)}, \omega_3^{(t_0)})$, $\gamma^{(t_0)} = (\Phi^{(t_0)}, \Theta^{(t_0)}, \Psi^{(t_0)})$ in time t_0, and a set of data, describing the torques that are acting on the model in time $t_0 \leq t \leq t_n$ and the moments of inertia. Using the Adams-Moulton predictor-corrector method, the simulator finds the state $\omega^{(t_n)}, \gamma^{(t_n)}$ of the model in time t_n.

In (A.1), an analytical solution of the Euler equations, for a specific case ($I_y = I_z = I$ and $L = 0$) was calculated.

Applying a different set of data to the simulator, for the special case I (section A.1.1), and comparing the output of the simulator with the correct results from the analytical solution (see page 169), the tables (A.1),(A.2),(A.3),(A.4) are obtained.

Applying different set of data to the simulator, for the special case II (section A.1.2), and comparing the output of the simulator, with the correct results from the analytical solution (see page 169), produces table (A.5). In this case $M = \sin \alpha t, N = \cos \alpha t$.

In all of these cases, $L = 0, I_y = I_z = I$. Observation of the tables show that the simulator is very accurate, assuming that the "step" value h is small enough (in these cases 0.01). Cases in which the simulator doesn't behave

A.2. COMPUTING A SOLUTION FOR EULER EQUATIONS

M	N	I_x	I	Solution ω_2	ω_3	Simulator $h = 0.1$ ω_2	ω_3
0.1	0.2	0.35	0.5	0.392864	0.652018	0.392864	0.652018
0.5	1.2	0.35	0.5	1.137075	2.442080	1.137075	2.442080
0.5	1.2	2.35	1.5	0.465961	1.037142	0.465961	1.037142
3.5	4.2	2.35	1.5	2.219290	2.882252	2.219290	2.882252
3.5	4.2	8.5	1.5	1.664537	2.828836	1.664537	2.828837

Table A.1: Comparison of simulator output and the analytical solution of Euler equations for a special case (I). $t_0 = 0.1, t_n = 1.0, w_1(t_0) = 0.1, w_2(t_0) = 0.2, w_3(t_0) = 0.3$

M	N	I_x	I	Solution ω_2	ω_3	Simulator $h = 0.01$ ω_2	ω_3
0.1	0.2	0.35	0.5	0.855007	1.414810	0.855007	1.414810
0.5	1.2	0.35	0.5	3.424223	7.106640	3.424223	7.106640
0.5	1.2	2.35	1.5	0.920351	2.717498	0.920351	2.717498
3.5	4.2	2.35	1.5	6.218786	8.966924	6.218786	8.966924
3.5	4.2	8.5	1.5	1.593256	1.549395	1.593256	1.549395

Table A.2: Comparison of simulator output and the analytical solution of Euler equations for a special case (I). $t_0 = 0.1, t_n = 3.0, w_1(t_0) = 0.1, w_2(t_0) = 0.2, w_3(t_0) = 0.3$

M	N	I_x	I	Solution ω_2	ω_3	Simulator $h = 0.01$ ω_2	ω_3
0.1	0.2	0.35	0.5	3.515710	0.393128	3.515710	0.393128
0.5	1.2	0.35	0.5	8.199056	4.090511	8.199056	4.090511
0.5	1.2	2.35	1.5	-3.611617	2.364514	-3.611617	2.364514
3.5	4.2	2.35	1.5	-4.993779	9.161920	-4.993779	9.161920
3.5	4.2	8.5	1.5	1.258575	-3.026339	1.258574	-3.026339

Table A.3: Comparison of simulator output and the analytical solution of Euler equations for a special case (I). $t_0 = 0.1, t_n = 3.0, w_1(t_0) = 1.2, w_2(t_0) = 1.5, w_3(t_0) = 2.0$

APPENDIX A. EULER EQUATIONS SOLUTIONS

M	N	I_x	I	Solution		Simulator $h = 0.1$	
				ω_2	ω_3	ω_2	ω_3
0.1	0.2	0.35	0.5	3.515710	0.393128	3.480007	0.479060
0.5	1.2	0.35	0.5	8.199056	4.090511	7.950859	4.090511
0.5	1.2	2.35	1.5	-3.611617	2.364514	-3.478635	2.525642
3.5	4.2	2.35	1.5	-4.993779	9.161920	-4.602382	9.208208
3.5	4.2	8.5	1.5	1.258575	-3.026339	1.477273	-2.956945

Table A.4: Comparison of simulator output and the analytical solution of Euler equations for a special case (I). $t_0 = 0.1, t_n = 3.0, w_1(t_0) = 1.2, w_2(t_0) = 1.5, w_3(t_0) = 2.0$

I_x	I	Solution		Simulator $h = 0.01$	
		ω_2	ω_3	ω_2	ω_3
0.35	0.5	-10.94997	-18.96509	-10.94997	-18.96509
2.35	1.5	-2.777501	8.190794	-2.777501	8.190794
8.5	1.5	5.901229	5.180303	5.901213	5.180375

Table A.5: Comparison of simulator output and the analytical solution of Euler equations for a special case (II). $t_0 = 0.1, t_n = 10.0, w_1(t_0) = 1.2, w_2(t_0) = 1.5, w_3(t_0) = 2.0$

A.2. COMPUTING A SOLUTION FOR EULER EQUATIONS

well are the cases in which h is not small enough and the initial value of $\omega(t_0) = (\omega_1(t_0), \omega_2(t_0), \omega_3(t_0))$. But even in this case, if the value of h is changed from 0.1 to 0.01 the model becomes very accurate (compare tables A.3, A.4).

Appendix B

An Attitude Control Simulator

We were orbiting above a ring—at almost zero thrust, in other words. I knew that once the booster fired—and it had to, since the starter was on—we'd get a side reaction and start tumbling.
Stanislaw Lem
(*More Tales of Pirx the Pilot*)

The ANSI C code of an attitude control simulator is given.

```
#include <stdio.h>
#include <math.h>

#define TOTAL_TIME      5.0
#define TIMESTEP        0.01
#define MAXITERATIONS   100001


int n = 0, n1 = 0 ;
double I1=1200, I2=22000, I3=25000, h, h1;
double   x[MAXITERATIONS], y[MAXITERATIONS], z[MAXITERATIONS];
double   Psi[MAXITERATIONS], Theta[MAXITERATIONS];
double   Phi[MAXITERATIONS];
double Phi0, Theta0, Psi0;
```

```c
double L, M, N;
double x0, y01, z0, t0, tn=0.0, tfinal;
double two_pi;

double f(double t, double w1, double w2, double w3)
{
        double temp;

        temp = (L + (I2 - I3)*w2*w3)/I1;
        return temp;
}


double g(double t, double w1, double w2, double w3)
{
        double temp;

        temp = (M + (I3 - I1)*w1*w3)/I2;
        return temp;
}


double v(double t, double w1, double w2, double w3)
{
        double temp;

        temp = (N + (I1 - I2)*w1*w2)/I3;
        return temp;
}


void Runge(double t0, double x0, double y0, double z0,double tn,
           double h)
{
        int times , i ;
        double kn1,kn2,kn3,kn4,ln1,ln2,ln3,ln4,mn1,mn2,mn3,mn4;

        x[n] = x0;
        y[n] = y0;
        z[n] = z0;
```

APPENDIX B. AN ATTITUDE CONTROL SIMULATOR

```
            times = (tn-t0)/h;   /* compute number of iterations */
            for (i = 0 ; i < times ; i++) {
                kn1 = f(t0+n*h,x[n],y[n],z[n]);
                ln1 = g(t0+n*h,x[n],y[n],z[n]);
                mn1 = v(t0+n*h,x[n],y[n],z[n]);

                kn2 = f(t0+n*h+1.0/2.0*h , x[n]+1.0/2.0*h*kn1 ,
                        y[n]+1.0/2.0*h*ln1 , z[n]+1.0/2.0*h*mn1);
                ln2 = g(t0+n*h+1.0/2.0*h , x[n]+1.0/2.0*h*kn1 ,
                        y[n]+1.0/2.0*h*ln1 , z[n]+1.0/2.0*h*mn1);
                mn2 = v(t0+n*h+1.0/2.0*h , x[n]+1.0/2.0*h*kn1 ,
                        y[n]+1.0/2.0*h*ln1 , z[n]+1.0/2.0*h*mn1);
                kn3 = f(t0+n*h+1.0/2.0*h , x[n]+1.0/2.0*h*kn2 ,
                        y[n]+1.0/2.0*h*ln2 , z[n]+1.0/2.0*h*mn2);
                ln3 = g(t0+n*h+1.0/2.0*h , x[n]+1.0/2.0*h*kn2 ,
                        y[n]+1.0/2.0*h*ln2 , z[n]+1.0/2.0*h*mn2);
                mn3 = v(t0+n*h+1.0/2.0*h , x[n]+1.0/2.0*h*kn2 ,
                        y[n]+1.0/2.0*h*ln2 , z[n]+1.0/2.0*h*mn2);
                kn4 = f(t0+n*h+h , x[n]+h*kn3 , y[n]+h*ln3 ,
                        z[n]+h*mn3);
                ln4 = g(t0+n*h+h , x[n]+h*kn3 , y[n]+h*ln3 ,
                        z[n]+h*mn3);
                mn4 = v(t0+n*h+h , x[n]+h*kn3 , y[n]+h*ln3 ,
                        z[n]+h*mn3);

                /* compute new values of x , y , z for step n+1 */

                x[n + 1] = x[n] + 1.0/6.0*h*(kn1 + 2.0*kn2 +
                           2.0*kn3 + kn4);
                y[n + 1] = y[n] + 1.0/6.0*h*(ln1 + 2.0*ln2 +
                           2.0*ln3 + ln4);
                z[n + 1] = z[n] + 1.0/6.0*h*(mn1 + 2.0*mn2 +
                           2.0*mn3 + mn4);
                ++n;
            }
    }
```

```
double f1(double t, double theta, double phi, double psi)
{
    double temp;
    int index;

    index = (t - t0)/h;
    temp = y[index]*cos(phi) - z[index]*sin(phi);
    return temp;
}

double g1(double t, double theta, double phi, double psi)
{
    double temp;
    int index;

    index = (t - t0)/h;
    temp = x[index] + y[index]*sin(phi)*tan(theta) +
           z[index]*cos(phi)*tan(theta);
    return temp;
}

double v1(double t, double theta, double phi, double psi)
{
    double temp;
    int index;

    index = (t - t0)/h;
    temp = (y[index]*sin(phi) + z[index]*cos(phi))/cos(theta);
    return temp;
}

void Runge1(double t0, double Theta0, double Phi0, double Psi0,
            double tn, double h)
{
    int times , i , round_temp;
    double kn1,kn2,kn3,kn4,ln1,ln2,ln3,ln4,mn1,mn2,mn3,mn4 ;
```

APPENDIX B. AN ATTITUDE CONTROL SIMULATOR

```
Theta[n1] = Theta0;
Phi[n1] = Phi0;
Psi[n1] = Psi0;
times = (tn-t0)/h;          /* compute number of iterations */
for (i = 0 ; i < times ; i++) {
    kn1 = f1(t0+n1*h,Theta[n1],Phi[n1],Psi[n1]);
    ln1 = g1(t0+n1*h,Theta[n1],Phi[n1],Psi[n1]);
    mn1 = v1(t0+n1*h,Theta[n1],Phi[n1],Psi[n1]);

    kn2 = f1(t0+n1*h+1.0/2.0*h , Theta[n1]+1.0/2.0*h*kn1 ,
        Phi[n1]+1.0/2.0*h*ln1 , Psi[n1]+1.0/2.0*h*mn1);
    ln2 = g1(t0+n1*h+1.0/2.0*h , Theta[n1]+1.0/2.0*h*kn1 ,
        Phi[n1]+1.0/2.0*h*ln1 , Psi[n1]+1.0/2.0*h*mn1);
    mn2 = v1(t0+n1*h+1.0/2.0*h , Theta[n1]+1.0/2.0*h*kn1 ,
        Phi[n1]+1.0/2.0*h*ln1 , Psi[n1]+1.0/2.0*h*mn1);
    kn3 = f1(t0+n1*h+1.0/2.0*h , Theta[n1]+1.0/2.0*h*kn2 ,
        Phi[n1]+1.0/2.0*h*ln2 , Psi[n1]+1.0/2.0*h*mn2);
    ln3 = g1(t0+n1*h+1.0/2.0*h , Theta[n1]+1.0/2.0*h*kn2 ,
        Phi[n1]+1.0/2.0*h*ln2 , Psi[n1]+1.0/2.0*h*mn2);
    mn3 = v1(t0+n1*h+1.0/2.0*h , Theta[n1]+1.0/2.0*h*kn2 ,
        Phi[n1]+1.0/2.0*h*ln2 , Psi[n1]+1.0/2.0*h*mn2);
    kn4 = f1(t0+n1*h+h , Theta[n1]+h*kn3 , Phi[n1]+h*ln3 ,
        Psi[n1]+h*mn3);
    ln4 = g1(t0+n1*h+h , Theta[n1]+h*kn3 , Phi[n1]+h*ln3 ,
        Psi[n1]+h*mn3);
    mn4 = v1(t0+n1*h+h , Theta[n1]+h*kn3 , Phi[n1]+h*ln3 ,
        Psi[n1]+h*mn3);

        /* compute new values of x , y , z for step n+1 */

    Theta[n1 + 1] = Theta[n1] + 1.0/6.0*h*(kn1 + 2.0*kn2 +
                2.0*kn3 + kn4);
    Phi[n1 + 1] = Phi[n1] + 1.0/6.0*h*(ln1 + 2.0*ln2 +
                2.0*ln3 + ln4);
    Psi[n1 + 1] = Psi[n1] + 1.0/6.0*h*(mn1 + 2.0*mn2 +
                2.0*mn3 + mn4);
    ++n1;
```

```c
        /* Modulo 2 pi angles */
        if (Psi[n1] > two_pi) {
            round_temp = Psi[n1]/two_pi;
            Psi[n1] -= two_pi*round_temp;
        }
    }
}

main()
{
    h = 0.01;
    h1 = 2.0*h;
    two_pi = 2.0*acos(-1.0);
    tn = TOTAL_TIME;

    /* initial conditions */

    x0 = 1.0;
    y01 = 2.0;
    z0 = 1.0;
    Theta0 = 1.2;
    Phi0 = 1.0;
    Psi0 = 3.0;

    /* plant left free, so no control inputs, all zero */

    L = 0;
    M = 0;
    N = 0;

    n = 0;
    Runge(t0 , x0 , y01 , z0 , tn , h);
    n1 = 0;
    Runge1(t0, Theta0 , Phi0 , Psi0 , tn , h1);

    printf("\nAt time %lf:\n\n", tn);
```

```
    printf("omega1 = %lf    omega2 = %lf    omega3 = %lf\n", x[n],
           y[n], z[n]);
    printf("theta = %lf     phi = %lf    psi = %lf\n", Theta[n1],
           Phi[n1],Psi[n1]);
}
```

Bibliography

[1] Aarts, E. and Korst, J. *Simulation Annealing and Boltzmann Machines: A Stochastic Approach to Combinatorial Optimization and Neural Computing.* John Wiley, 1990.

[2] Ackley, D. H., Hinton, G. E., and Sejnowski, T. J. A learning algorithm for Boltzmann machines. *Cognitive Science*, (9):147–169, 1985.

[3] Alavi, F. and Taylor, J. G. A basis for long-range inhibition across cortex. In Sirosh, J., Mikkulainen, R., and Choe, Y., editors, *Lateral Interactions in the Cortex*. Electronic book, http://www.cs.utexas.edu/users/nn/web-pubs/htmlbook96, 1995.

[4] Aleksander, I. and Morton, H. *An Introduction to Neural Computing.* International Thomson Computer Press, second edition, 1995.

[5] Allman, W. F. *Apprentices of wonder-Inside the Neural Network.* Bantam Books, 1989.

[6] Amit, D. *Modeling Brain Function.* Cambridge University Press, 1989.

[7] Anderson, C. W. Learning to control an inverted pendulum using neural networks. *IEEE Control Systems Magazine*, pages 31–37, April 1989.

[8] Anderson, J. A. and Rosenfeld, E., editors. *Neurocomputing: Foundations of Research.* MIT Press, 1990.

[9] Anderson, J. D. *Introduction to flight.* McGraw-Hill, 1978.

[10] Arnold, V. I. *Mathematical Methods of Classical Mechanics.* Springer-Verlag, 1978.

[11] Ashby, W. R. *Design for a brain.* Chapman and Hall, 1960.

[12] Astrom, K. J. Towards intelligent control. *IEEE Control Systems Magazine*, April 1988.

[13] Astrom, K. J. and Wittenmark, B. *Adaptive Control*. Addison-Wesley, first edition, 1989.

[14] Baker, G. and Gollub, J. *Chaotic Dynamics - An Introduction*. Cambridge University Press, 1990.

[15] Baker, J. E. Reducing bias and inefficiency in the selection algorithm. In Grefenstette, J. J., editor, *Genetic Algorithms and their Applications: Proceedings of the second international conference on genetic algorithms*. Lawrence Erlbaum Associates, 1987.

[16] Barron, A. R. Universal approximation bounds for superpositions of a sigmoidal function. *IEEE Transactions on Information Theory*, 39(3):930–945, 1993.

[17] Barto, A. G., Sutton, R. S., and Anderson, C. W. Neuronlike adaptive elements that can solve difficult control problems. *IEEE Transactions on Systems, Man and Cybernetics*, SMC-13:834–846, 1983.

[18] Bavarian, B. Introduction to neural networks for intelligent control. *IEEE Control Systems Magazine*, pages 3–7, April 1988.

[19] Bender, C. M. and Orszag, S. A. *Advanced Mathematical Methods for Scientists and Engineers*. McGraw-Hill, 1978.

[20] Bertsekas, D. P. and Tsitsiklis, J. N. *Neuro-Dynamic Programming*. Athena Scientific, 1996.

[21] Bloch, A. M., Krishnaprasad, P. S., Marseden, J. E., et al. Stabilization of rigid body dynamics by internal and external torques. *Automatica*, 28(4):745–756, 1992.

[22] Bondi, H. The rigid body dynamics of unidirectional spin. *Proc. R. Soc. Lond.*, A405:265–474, 1986.

[23] Bosch, P. P. J. V. D., Jongkind, W., and Swieten, A. C. M. V. Adaptive attitude control for large-angle slew manoeuvres. *Automatica*, 22(2):209–215, 1986.

[24] Bose, N. K. and Liang, P. *Neural Network Fundamentals with Graphs, Algorithms and Applications*. McGraw Hill, 1996.

[25] Boyce, W. E. and DiPrima, R. C. *Elementary Differential Equations and Boundary Value Problems*. John Wiley and Sons, New York, fifth edition, 1992.

[26] Branets, V. N., Weinberg, D. M., Verestchagin, V. P., et al. Development experience of the attitude control system using single-axis control moment gyros for long-term orbiting space stations. *ACTA Astronautica*, 18:91–98, 1988.

[27] Brooks, V. B. *The Neural Basis of Motor Control*. Oxford University Press, 1986.

[28] Broomhead, D. S. and Lowe, D. Multivariable functional interpolation and adaptive networks. *Complex Systems*, 2:321–355, 1988.

[29] Bryson, A. E. and Ho, Y. *Applied Optimal Control*. John-Wiley, 1975.

[30] Bulsari, A. Some analytical solutions to the general approximation problem for feedforward neural networks. *Neural Networks*, 6(6), November 1993.

[31] Campbell, D. Introduction to nonlinear phenomena. In Stein, D. L., editor, *Lectures in the Sciences of Complexity*. Addison Wesley, 1989.

[32] Carpenter, G. A. Distributed ART networks for learning, recognition, and prediction. In *World Congress on Neural Networks, San Diego, California*, pages 333–344. Lawrence Erlbaum Associates,Inc. and INNS Press, September 1996.

[33] Carpenter, G. A. and Grossberg, S. ART 2: self-organization of stable category recognition codes for analog input patterns. *Applied Optics*, 26(23):4919–4930, 1987.

[34] Carpenter, G. A. and Grossberg, S. A massively parallel architecture for a self-organising neural pattern recognition machine. *Computer Vision, Graphics, and Image Processing*, 37:54–115, 1987.

[35] Carpenter, G. A. and Grossberg, S. The ART of adaptive pattern recognition by a self-organizing neural network. *Computer*, pages 77–88, March 1988.

[36] Carpenter, G. A. and Grossberg, S. Adaptive resonance theory: Neural network architectures for self-organizing pattern recognition. In Eckmiller, R., Hartmann, G., and Hauske, G., editors, *Parallel Processing in Neural Systems and Computers*, pages 383–389. Elsevier Science Publishers B.V (North Holland), 1990.

[37] Carpenter, G. A., Grossberg, S., and Doerschuk, P. I. Feedback models. In Fiesler, E. and Beale, R., editors, *Handbook of Neural Computation*. IOP Publishing Ltd and Oxford University Press, 1997.

[38] Ceballos, D. C. Compensating structure and parameter optimization for attitude control of a flexible spacecraft. *Journal of Guidance, Control and Dynamics*, 9(2):248–249, March-April 1986.

[39] Chavy, S. and Fantar, F. Design of attitude control system for injection of spinned payloads. *ACTA Astronautica*, 25(4):185–197, 1991.

[40] Chobotov, V. A. *Spacecraft Attitude Dynamics and Control*. Krieger Publishing Company, 1991.

[41] Chretien, J. P., Reboulet, C., and Rodrigo, P. Attitude control of a satellite with a rotating solar array. *Journal of Guidance, Control and Dynamics*, 5(6):589–596, November-December 1982.

[42] Conte, S. D. and deBoor, C. *Elementary Numerical Analysis : An Algorithmic Approach*. McGraw-Hill, New York, third edition, 1980.

[43] Courant, R. *Differential and Integral Calculus*. Interscience Publishers, New York, second edition, 1937.

[44] Crouch, P. E. Spacecraft attitude control and stabilization: Applications of geometric control theory to rigid body models. *IEEE Transactions on Automatic Control*, AC-29:321–331, 1984.

[45] Cybenco, G. Approximation by superpositions of a sigmoidal function. *Mathematics of Control, Signals and Systems*, 2, 1989.

[46] Daley, S. and Gill, K. F. Attitude control of a spacecraft using an extended self-organizing fuzzy logic controller. *Proceedings of the Institute of Mechanical Engineers*, 201(C2):97–106, 1987.

[47] Darwin, C. *The Origin of Species by Means of Natural Selection*. Penguin, 1985. republication.

BIBLIOGRAPHY

[48] Davis, L., editor. *Genetic Algorithms and Simulated Annealing*. Morgan Kaufmann, 1987.

[49] Davis, L., editor. *Handbook of Genetic Algorithms*. Van Nostrand Reinhold, 1991.

[50] Davis, W. R. and Levinson, D. A. Attitude control of a satellite containing a spinning rotor with a nadir-pointing axis. *Journal of the Astronautical Sciences*, 38(1):1–19, January-March 1990.

[51] De, P. K. and Iyer, A. Stable fraction approach and attitude control of a spinning satellite using a gyrotorquer. *International Journal of Systems Science*, 23(3):311–328, 1992.

[52] Denker, J. S., editor. *Neural Networks for Computation*, number 151 in AIP Conference Proceedings, New York, 1986. American Institute of Physics.

[53] Dennis, J. E. and Schnabel, R. B. *Numerical Methods for Unconstrained Optimization and Nonlinear Equations*. Prentice Hall, 1983.

[54] Dettleff, G., Boettcher, R. D., Dankert, C., et al. Attitude control thruster plume flow modeling and experiments. *Journal of Spacecraft and Rockets*, 23(5):476–481, September-October 1986.

[55] Ditto, W. L. and Pecora, L. M. Mastering chaos. *Scientific American*, August 1993.

[56] Dodds, S. J. Adaptive, high precision, satellite attitude control for microprocessor implementation. *Automatica*, 17(4):563–574, 1981.

[57] Dodds, S. J. and Walker, A. B. Sliding-mode control system for the three-axis attitude control of rigid-body spacecraft with unknown dynamics parameters. *International Journal of Control*, 54(4):737–761, 1991.

[58] Dodds, S. J. and Williamson, S. E. A signed switching time bang-bang attitude control law for fine pointing of flexible spacecraft. *International Journal of Control*, 40(4):795–811, 1984.

[59] Dracopoulos, D. C. *Neuromodelling, Adaptive Neurocontrol and the Attitude Control Problem*. Ph.D. thesis, Imperial College of Science, Technology and Medicine, University of London, London SW7 2BZ, United Kingdom, March 1994.

[60] Dracopoulos, D. C. Evolutionary control of a satellite. In Koza, J. R., Kalyanmoy, D., Dorigo, M., et al., editors, *Genetic Programming 1997: Proceedings of the Second Annual Conference*, Stanford University, San Francisco, CA, July 13–16 1997. Morgan Kaufmann.

[61] Dracopoulos, D. C. Genetic algorithms and genetic programming in control engineering. In Dasgupta, D. and Michalewicz, Z., editors, *Evolutionary Algorithms in Engineering Applications*. Springer Verlag, 1997.

[62] Dracopoulos, D. C. and Jones, A. J. Modeling dynamic systems. In *1st World Congress on Neural Networks Proceedings*, pages 289–292. INNS/Erlbaum Press, 1993.

[63] Dracopoulos, D. C. and Jones, A. J. Neuromodels of analytic dynamic systems. *Neural Computing & Applications*, 1(4):268–279, 1993.

[64] Dracopoulos, D. C. and Jones, A. J. An adaptive neurocontrol design applied to the attitude control problem. In *Applications of Neural Networks V, OE/Aerospace Sensing, SPIE*, pages 380–391, 1994.

[65] Dracopoulos, D. C. and Jones, A. J. Neural networks and genetic algorithms for the attitude control problem. In *Neuronîmes 94*, 1994.

[66] Dracopoulos, D. C. and Jones, A. J. Neuro-genetic adaptive attitude control. *Neural Computing & Applications*, 2(4):183–204, 1994.

[67] Etkin, B. *Dynamics of Flight - Stability and Control*. John Wiley and Sons, second edition, 1982.

[68] Farrell, J. L. *Integrated Aircraft Navigation*. Academic Press, New York, 1976.

[69] Fenton, J. and Gill, K. F. Flexible spacecraft attitude control using a simple p+d algorithm. *Aeronautical Journal*, 85(844):185–189, May 1981.

[70] Fraser, A. M. and Swinney, H. L. Independent coordinates for strange attractors from mutual information. *Physical Review A*, 33(2):1134–1140, 1986.

[71] Furuta, K., Ochiai, T., and Ono, N. Attitude control of a triple inverted pendulum. *International Journal of Control*, 39(6):1351–1365, 1984.

[72] Gleick, J. *Chaos*. Cardinal, 1987.

[73] Goldberg, D. E. *Genetic Algorithms in Search, Optimization and Machine Learning.* Addison Wesley, 1989.

[74] Goldstein, H. *Classical Mechanics.* Addison Wesley, second edition, 1980.

[75] Gollub, G. H. and Loan, C. F. V. *Matrix Computations.* John Hopkins University Press, second edition, 1989.

[76] Goodwin, G. C. and Sin, K. S. *Adaptive Filtering Prediction and Control.* Prentice Hall, 1984.

[77] Gorges-Schleuter, M. *Genetic Algorithms and Population Structures:A massively parallel algorithm.* Ph.D. thesis, University of Dortmund, 1990.

[78] Grefenstette, J. and Gopal, R. Genetic algorithms for the travelling salesman problem. In J.D.Schaffer, editor, *Proceedings of the third international conference on genetic algorithms.* Morgan Kaufmann, 1989.

[79] Grefenstette, J. J. and Baker, J. E. How genetic algorithms work: A critical look at implicit parallelism. In J.D.Schaffer, editor, *Proceedings of the third international conference on genetic algorithms.* Morgan Kaufmann, 1989.

[80] Greffenstette, J. J. Optimization of control parameters for genetic algorithms. *IEEE Transactions on Systems, Man and Cybernetics*, SMC-16:122–128, 1986.

[81] Gregory, J. C. and Peters, P. N. Measurement of the passive attitude control performance of a recovered spacecraft. *Journal of Guidance, Control and Dynamics*, 15(1):282–284, January-February 1992.

[82] Guckenheimer, J. and Holmes, P. *Nonlinear Oscillations, Dynamical Systems, and Bifurcations of Vector Fields.* Springer Verlag, first edition, 1983.

[83] Guez, A., Eilbert, J. L., and Kam, M. Neural network architecture for control. *IEEE Control Systems Magazine*, April 1988.

[84] Haykin, S. *Neural Networks: A Comprehensive Foundation.* MacMillan, 1994.

[85] Hebb, D. O. *The Organization of Behavior.* Wiley, New York, 1949.

[86] Hecht-Nielsen, R. *Neurocomputing*. Addison Wesley, 1990.

[87] Herman, A. L. and Conway, B. A. Optimal spacecraft attitude control using collocation and nonlinear programming. *Journal of Guidance, Control and Dynamics*, 15(5):1287–1289, 1991.

[88] Hermes, H. On a stabilizing feedback attitude control. *Journal of Optimization Theory and Applications*, 31(3):373–384, July 1980.

[89] Hermes, H. and Hogenson, D. The explicit synthesis of stabilizing (time optimal) feedback controls for the attitude control of a rotating satellite. *Applied Mathematics and Computation*, 16:229–240, 1985.

[90] Hertz, J., Kroph, A., and Palmer, R. G. *Introduction to the theory of Neural Computation*. Addison Wesley, 1991.

[91] Hess, R. A. Analysis of aircraft attitude control systems prone to pilot-induced oscillations. *Journal of Guidance, Control and Dynamics*, 7(1):106–112, January-February 1984.

[92] Hildebrand, F. B. *Introduction to Numerical Analysis*. Dover, 1974.

[93] Hillis, W. D. *The Connection Machine*. MIT Press, 1985.

[94] Hinton, G. E. and Sejnowski, T. J. Optimal perceptual inference. pages 448–453. Proceedings of the IEEE Conference on Computer Vision and Pattern Recognition, 1983.

[95] Hinton, G. E. and Sejnowski, T. J. Learning and relearning in Boltzmann machines. In Rumelhart, D., McClelland, J., and the PDP research group, editors, *Parallel Distributed Processing*, volume 1, chapter 7. MIT Press, 1986.

[96] Hodgart, M. S. Attitude control and dynamics of uosat angular motion. *Radio and Electronic Engineer*, 52(8/9):379–384, August-September 1982.

[97] Holland, J. H. *Adaptation in Natural and Artificial Systems*. The University of Michigan Press, 1975.

[98] Holland, J. H., Holyoak, K. J., Nisbett, R. E., et al. *Induction: Processes of Inference, Learning and Discovery*. MIT Press, 1986.

[99] Hopfield, J. J. Neural networks and physical systems with emergent collective computational abilities. Number 79, pages 2554–2558. Proceedings of the National Academy of Sciences, 1982.

[100] Hopfield, J. J. Neurons with graded response have collective computational properties like those of two-state neurons. Number 81, pages 3088–3092. Proceedings of the National Academy of Sciences, 1984.

[101] Hornik, K. Multilayer feedforward networks are universal approximators. *Neural Networks*, 2:359–366, 1989.

[102] Humble, R. W. Two dimensional tethered satellite attitude dynamics. *Journal of the Astronautical Sciences*, 38(1):21–27, January-March 1990.

[103] Iyer, A. and Singh, S. N. MFDs of spinning satellite and attitude control using gyrotorquers. *IEEE Transactions on Aerospace and Electronic Systems*, 25(5):611–620, September 1989.

[104] Jahangir, E. and Howe, R. M. Time-optimal control scheme for a spinning missile. *Journal of Guidance, Control and Dynamics*, 16(2):346–353, March-April 1993.

[105] Jahanshahi, M. H. Simultaneous calibrations of voyager celestial and inertial attitude control systems in flight. *IEEE Transactions on Aerospace and Electronic Systems*, AES-18(1):21–28, January 1982.

[106] J.M.T.Thompson and H.B.Stewart. *Nonlinear Dynamics and Chaos*. John Wiley and Sons, 1986.

[107] Jones, A. J. A schema theorem for trees. personal communication.

[108] Jones, A. J. Models of living systems: Evolution and neurology. Department of Computer Science, University of Wales, Cardiff, 1997. Forthcoming book publication.

[109] Jong, K. A. D., Spears, W. M., and Gordon, D. F. Using genetic algorithms for concept learning. *Machine Learning*, 13:161–188, 1993.

[110] Jordan, M. I. and Rumelhart, D. E. Forward models: Supervised learning with a distal teacher. *Cognitive Science*, 16:307–354, 1992.

[111] Jr., R. A. C. and Yeake, G. S. Active attitude control of a spinning symmetrical satellite in an elliptic orbit. *Journal of Guidance, Control and Dynamics*, 6(4):315–320, July-August 1983.

[112] Junkins, J. L., Rajaram, S., Baracat, W. A., et al. Precision autonomous satellite attitude control using momentum transfer and magnetic torquer. *Journal of the Astronautical Sciences*, XXX(1):31–48, January-March 1982.

[113] Katayama, M. and Kawato, M. Learning trajectory and force control of an artificial muscle arm by parallel-hierarchical neural network model. In *Neural Information Processing Systems 3*. Morgan Kaufmann, 1991.

[114] Kawato, M. Computational schemes and neural network models for formation and control of multijoint arm trajectory. In Miller, S. and Werbos, editors, *Neural Networks for Control*. MIT Press, 1990.

[115] Kawato, M., Uno, Y., Isobe, M., et al. Hierarchical neural network model for voluntary movement with application. *IEEE Control Systems Magazine*, April 1988.

[116] Keller, H. B. *Numerical Methods for Two-Point Boundary Value Problems*. Blaisdell Publishing Company, 1968.

[117] Kirkpatrick, S., Jr., C. D. G., and Vecchi, M. P. Optimization by simulation annealing. *Science*, 220(4598):671–680, 1983.

[118] Kline-Schoder, R. J. and Powell, J. D. Precision attitude control for tethered satellites. *Journal of Guidance, Control and Dynamics*, 16(1):168–174, January-February 1993.

[119] Kohonen, T. *Self Organisation and Associative Memory*. Springer-Verlag, second edition, 1988.

[120] Koza, J. R. *Genetic Programming*. MIT Press, 1982.

[121] Koza, J. R. *Genetic Programming II*. MIT Press, 1994.

[122] K.Y.Goldberg and B.A.Pearlmutter. Using backpropagation with temporal windows to learn the dynamics of the cmu direct drive arm ii. In *Neural Information Processing Systems 1*. Morgan Kaufmann, 1989.

[123] Lampert, J. D. *Computational Methods In Ordinary Differential Equations*. John Wiley and Sons, 1973.

[124] Lapedes, A. and Farber, R. How neural nets work. In *Proceedings of IEEE, Denver Conference on Neural Nets*, 1987.

[125] Lapedes, A. and Farber, R. Nonlinear signal processing using neural networks: prediction and system modelling. Technical Report LA-UR-87-2662, Los Alamos National Laboratory, 1987.

[126] Leipnik, R. B. and Newton, T. A. Double strange attractors in rigid body motion with linear feedback control. *Physics Letters*, 86A:63–67, 1981.

[127] Lemke, L. G., Powell, J. D., and He, X. Attitude control of tethered spacecraft. *Journal of the Astronautical Sciences*, 35(1):41–55, January-March 1987.

[128] Levin, A. U. Recursive identification using feedforward networks. personal communication.

[129] Lin, Y. Y. and Kraige, L. G. Enhanced techniques for solving the two-point boundary-value problem associated with the optimal attitude control of spacecraft. *Journal of the Astronautical Sciences*, 37(1):1–15, 1989.

[130] Long, T. W. Satellite attitude control in the equivalent axis frame. *ACTA Astraunotica*, 26(5):299–305, 1992.

[131] Luenberger, D. G. *Introduction to Dynamic Systems*. John Wiley and Sons, 1979.

[132] Luenberger, D. G. *Linear and Nonlinear Programming*. Addison Wesley, second edition, 1984.

[133] Madany, M. M. E. and Abduljabbar, Z. On the ride and attitude control of road vehicles. *Computers and Structures*, 42(2):245–253, 1992.

[134] Mandelbrot, B. *The fractal Geometry of Nature*. Freeman, 1982.

[135] McClelland, J. and Rumelhart, D. *Explorations in Parallel Distributed Processing - A Handbook of Models, Programs and Exercises*. MIT Press, 1988.

[136] Melo, J. D. D. and Singh, S. N. Vtol aircraft control output tracking sensitivity design. *IEEE Transactions on Aerospace and Electronic Systems*, AES-20(2):82–93, March 1984.

[137] Mendel, J. M. and McLaren, R. W. Reinforcement learning control and pattern recognition systems. In Press, A., editor, *Adaptive Learning and Pattern Recognition Systems*, chapter 8. J. M. Mendel and K. S. Fu, 1970.

[138] Metropolis, N., Rosenbluth, A., Rosenbluth, M., et al. Equation of state calculations by fast computing machines. *Journal of Chemical Physics*, 1953.

[139] Meyer, G. On the use of Euler's theorem on rotations for the synthesis of attitude control systems. Technical Report TN D-3643, NASA, 1966.

[140] Meyer, G. Design and global analysis of spacecraft attitude control systems. Technical Report TR R-361, NASA, 1971.

[141] Miller, W., Sutton, R. S., and Werbos, P. J., editors. *Neural Networks for Control*. MIT Press, 1990.

[142] Minsky, M. and Papert, S. *Perceptrons-An Introduction to Computational Geometry*. MIT Press, 1969.

[143] Mohler, R. R. *Nonlinear systems: Dynamics and Control*. Prentice Hall, 1991.

[144] Moody, J. and Darken, C. Learning with localized receptive fields. In Touretzky, D., Hinton, G., and Sejnowski, T., editors, *Proceedings of the 1988 Connectionist Summer School*. Morgan Kaufmann, 1988.

[145] Moon, F. C. *Chaotic and Fractal Dynamics*. John Wiley and Sons, 1992.

[146] Muhlfelder, L. Evolution of an attitude control system for body-stabilized communication spacecraft. *Journal of Guidance, Control and Dynamics*, 9(1):108–112, January-February 1986.

[147] Narendra, K. S. Adaptive control of dynamical systems using neural networks. In White, D. A. and Sofge, D. A., editors, *Handbook of Intelligent Control*, chapter 5. Van Nostrand Reinhold, 1992.

[148] Narendra, K. S. and Annaswamy, A. M. *Stable Adaptive Systems*. Prentice Hall, 1989.

[149] Narendra, K. S. and Mukhopadhyay, S. Intelligent control using neural networks. *IEEE Control Systems Magazine*, pages 11–18, April 1992.

[150] Narendra, K. S. and Parthasarathy, K. Neural networks and dynamical systems, part I: A gradient approach to Hopfield networks. Technical Report 8820, Yale University, 1988.

[151] Narendra, K. S. and Parthasarathy, K. Neural networks and dynamical systems, part II: Identification. Technical Report 8902, Yale University, 1989.

[152] Narendra, K. S. and Parthasarathy, K. Neural networks and dynamical systems, part III: Control. Technical Report 8909, Yale University, 1989.

[153] Narendra, K. S. and Parthasarathy, K. Identification and control of dynamical systems using neural networks. *Neural Networks*, 1(1):4–27, 1990.

[154] Nicosia, S. and Tomei, P. Nonlinear observer and output feedback attitude control of spacecraft. *IEEE Transactions on Aerospace and Electronic Systems*, 28(4):970–977, 1992.

[155] Nilsson, N. J. *Learning Machines-Foundations of Trainable Pattern Classifying Systems*. McGraw-Hill, 1965.

[156] Ott, E., Grebogi, C., and Yorke, J. Controlling chaos. *Physical Review Letters*, 64(11), 1990.

[157] Packard, N. H., Crutchfield, J. P., Farmer, J. D., et al. Geometry from a time series. *Physical Review Letters*, 45(9):712–716, September 1980.

[158] Paielli, R. A. and Bach, R. E. Attitude control with realization of linear error dynamics. *Journal of Guidance, Control and Dynamics*, 16(1):182–189, January-February 1993.

[159] Pande, K. C. and Venkatachalam, R. Optimal solar pressure attitude control of spacecraft-i. *ACTA Astronautica*, 9(9), 1982.

[160] Pao, Y.-H. *Adaptive Pattern Recognition and Neural Networks*. Addison Wesley, 1989.

[161] Parkinson, B. W. and Kasdin, N. J. A magnetic attitude control system for precision pointing of the rolling GP-B spacecraft. *ACTA Astraunotica*, 21(6/7):477–486, 1990.

[162] Parks, P. C. Lyapunov redesign of model reference adaptive control systems. *IEEE Transactions on Automatic Control*, 11:362–367, 1966.

[163] Parlos, A. G. and Sunkel, J. W. Adaptive attitude control and momentum management for large-angle spacecraft maneuvers. *Journal of Guidance, Control and Dynamics*, 15(4):1018–1028, July-August 1992.

[164] Piper, G. E. and Kwatny, H. G. Complicated dynamics in spacecraft attitude control systems. *Journal of Guidance, Control and Dynamics*, 15(4):825–831, July-August 1992.

[165] Pope, T. Real-time simulation of the space station freedom attitude control system. *Simulation*, 57(1):17–25, 1991.

[166] Prokhorov, D. V. and II, D. C. W. Advanced adaptive critic designs. In *World Congress on Neural Networks, San Diego, California*, pages 83–87. Lawrence Erlbaum Associates,Inc. and INNS Press, September 1996.

[167] Prokhorov, D. V., Santiago, R., and II, D. C. W. Adaptive critic designs: A case study for neurocontrol. *Neural Networks*, 8(9):1367–1372, 1995.

[168] Psaltis, D., Sideris, A., and Yamamura, A. A. A multilayered neural network controller. *IEEE Control Systems Magazine*, pages 17–21, April 1988.

[169] Rahasingh, C. K. and Shrivastava, S. K. Orbit and attitude control of a geostationary inertially oriented large flexible plate-like spacecraft. *ACTA Astronautica*, 15(11):823–832, 1987.

[170] Rawlins, G. J. E., editor. *Foundations of Genetic Algorithms*. Morgan Kaufmann, 1991.

[171] Rhee, I. and Speyer, J. L. Robust momentum management and attitude control system for the space station. *Journal of Guidance, Control and Dynamics*, 15(2):342–351, March-April 1992.

[172] Routh, E. J. *Dynamics of a System of Rigid Bodies: Part II*. Macmillan and Co., 1892.

[173] Roux, J. C., Simoyi, R. H., and Swinney, H. L. Observations of a strange attractor. *Physica D*, 8:257–266, 1983.

[174] Rudin, W. *Principles of Mathematical Analysis*. McGraw-Hill, third edition, 1976.

[175] Rumelhart, D., McClelland, J., and the PDP research group. *Parallel Distributed Processing - Explorations in the Microstructure Cognition*, volume 1. MIT Press, 1986.

[176] Rumelhart, D., McClelland, J., and the PDP research group. *Parallel Distributed Processing - Explorations in the Microstructure Cognition*, volume 2. MIT Press, 1986.

[177] Schuster, H. G. *Deterministic Chaos*. Physik-Verlag, 1984.

[178] Seydel, R. *From Equilibrium to Chaos*. Elsevier, 1988.

[179] Shao-hua, Y. Satellite attitude control with decomposed controller. *Journal of Guidance, Control and Dynamics*, 4(6):584–585, November-December 1981.

[180] Sheela, B. V. and Shekhar, C. New star identification technique for attitude control. *Journal of Guidance, Control and Dynamics*, 14(2):477–480, March-April 1991.

[181] Shimada, I. and Nagashima, T. A numerical approach to ergodic problem of dissipative dynamical systems. *Progress of Theoretical Physics*, 61:1605–1616, 1979.

[182] Singh, S. N. Nonlinear attitude control of flexible spacecraft under disturbance torque. *ACTA Astronautica*, 13(8):507–514, 1986.

[183] Singh, S. N. Attitude control of a three rotor gyrostat in the presence of uncertainty. *Journal of the Astronautical Sciences*, 35(3):329–345, July-September 1987.

[184] Singh, S. N. Nonlinear adaptive attitude control of spacecraft. *IEEE Transactions on Aerospace and Electronic Systems*, AES-23(3):371–379, May 1987.

[185] Singh, S. N. Robust nonlinear attitude control of flexible spacecraft. *IEEE Transactions on Aerospace and Electronic Systems*, AES23(3):380–387, May 1987.

[186] Singh, S. N. Flexible spacecraft maneuver: Inverse attitude control and modal stabilization. *ACTA Astronautica*, 17(1):1–9, 1988.

[187] Singh, S. N. and Araújo, A. D. D. Asymptotic reproducibility in nonlinear systems and attitude control of gyrostat. *IEEE Transactions on Aerospace and Electronic Systems*, AES-20(2):94–103, March 1984.

[188] Singh, S. N. and Bossart, T. C. Exact feedback linearization and control of space station using cmg. *IEEE Transactions on Automatic Control*, 38(1):184–187, January 1993.

[189] Singh, S. N. and Iyer, A. Nonlinear decoupling sliding mode control and attitude control of spacecraft. *IEEE Transactions on Aerospace and Electronic Systems*, 25(5):621–633, September 1989.

[190] Skaar, S. B., Tang, L., and Yalda-Mooshabad, I. On-off attitude control of flexible satellites. *Journal of Guidance, Control and Dynamics*, 9(4):507–510, July-August 1986.

[191] Slotine, J.-J. E. and Li, W. *Applied Nonlinear Control*. Prentice Hall, 1991.

[192] Sparrow, C. T. *The Lorenz Equations: Bifurcations, Chaos and Strange Attractors*. Springer Verlag, 1982.

[193] Stewart, I. *Does God Play Dice? The New Mathematics of Chaos*. Penguin Books, 1989.

[194] Sutton, R. S., Barto, A. G., and Williams, R. J. Reinforcement learning is direct adaptive optimal control. *IEEE Control Systems Magazine*, pages 19–22, April 1992.

[195] Synge, J. L. and Griffith, B. *Principles of Mechanics*. McGraw-Hill, 1959.

[196] Syswerd, G. Uniform crossover in genetic algorithm. In J.D.Schaffer, editor, *Proceedings of the third international conference on genetic algorithms*. Morgan Kaufmann, 1989.

[197] Syswerda, G. A study of reproduction in generational and steady-state genetic algorithm. In *Foundations of Genetic Algorithms*, pages 94–101. Morgan Kaufman, 1991.

[198] Takens, F. Detecting strange attractors in turbulence. In Dold, A. and Eckmann, B., editors, *Dynamical Systems and Turbulence*, pages 366–381. Springer-Verlag, 1980.

[199] Taylor, J. G. The future of neural networks. In Fiesler, E. and Beale, R., editors, *Handbook of Neural Computation*. IOP Publishing Ltd and Oxford University Press, 1997.

[200] Tcuchiya, K., Inoue, M., Wakasuqi, N., et al. Advanced reaction wheel controller for attitude control of spacecraft. *ACTA Astronautica*, 9(12):697–702, 1982.

[201] Teichmann, F. K. *Fundamentals of Aircraft Flight*. Hayden Book Company Inc., 1974.

[202] Thomas, S. Dynamics of spacecrafts and manipulators. *Simulation*, 57(1):56–72, 1991.

[203] Touretzky, D., Hinton, G., and Sejnowski, T., editors. *Proceedings of the 1988 Connectionist Models Summer School*. Morgan Kauffmann, 1989.

[204] Vadali, R. S. and Oh, H. S. Space station attitude control and momentum management: A nonlinear look. *Journal of Guidance, Control and Dynamics*, 15(3):577–586, May-June 1992.

[205] Vanderplaats, G. N. *Numerical Optimization Techniques For Engineering Designs*. McGraw-Hill, 1984.

[206] Venkatachalam, R. and Pande, K. C. Optimal solar pressure attitude control of spacecraft-ii. *ACTA Astronautica*, 9(9):541–545, 1982.

[207] Wasserman, P. D. *Neural Computing - Theory and Practice*. Van Nostrand Reinhold, 1989.

[208] Wen, J. T.-Y. and Kreutz-Delgado, K. The attitude control problem. *IEEE Transactions on Automatic Control*, 36(10):1148–1162, October 1991.

[209] Werbos, P. Neurocontrol and related techniques. In Maren, A., editor, *Handbook of Neural Computing Applications*, pages 345–381. Academic Press, 1990.

[210] Werbos, P. Supervised learning: Can it escape its local minimum? In *First World Congress on Neural Networks*, July, 1993. INNS Press/Erlbaum.

[211] Werbos, P. J. Generalization of backpropagation with application to a recurrent gas market model. *Neural Networks*, 1:339–359, 1988.

[212] Werbos, P. J. Maximizing long-term gas industry profits in two minutes in lotus using neural network methods. *IEEE Transactions on Systems, Man and Cybernetics*, 19(2):315–333, March-April 1989.

[213] Werbos, P. J. Backpropagation through time: What it does and how to do it. *Proceedings of the IEEE*, 78(10), 1990.

[214] Werbos, P. J. An overview of neural networks for control. *IEEE Control Systems Magazine*, pages 40–41, January 1991.

[215] Werbos, P. J. Brain-like intelligence in artificial models: How can we really get there? *INNS Above Threshold*, 2(2), June 1993.

[216] Werbos, P. J. *The Roots of Backpropagation: From Ordered Derivatives to Neural Networks and Political Forecasting.* John Wiley and Sons, 1994.

[217] Werbos, P. J. Learning in the brain: An engineering interpretation. In Pribram, K., editor, *Learning as Self-Organization*. Erlbaum, 1996.

[218] Werbos, P. J. Neurocontrol, biological intelligence and engineering applications: Evaluation and prognosis. In *BMBF Proceedings on Neuroinformatiks and AI*, Berlin, 1996.

[219] Werbos, P. J. and Pang, X. Generalized maze navigation: Srn critics solve what feedforward or hebbian nets cannot. In *World Congress on Neural Networks, San Diego, California*, pages 88–93. Lawrence Erlbaum Associates,Inc. and INNS Press, September 1996.

[220] White, D. A. and Sofge, D. A., editors. *Handbook of Intelligent Control*. Van Nostrand Reinhold, 1992.

[221] Whitle, D. The genitor algorithm and selection pressure: Why rank based allocation of reproductive trials is best. In J.D.Schaffer, editor, *Proceedings of the third international conference on genetic algorithms*. Morgan Kaufmann, 1989.

[222] Whitley, D., Starkweather, T., and Fuquay, D. Scheduling problems and traveling salesmen: The genetic edge recombination operator. In J.D.Schaffer, editor, *Proceedings of the third international conference on genetic algorithms*. Morgan Kaufmann, 1989.

[223] Widrow, B. and Stearns, S. D. *Adaptive Signal Processing*. Prentice Hall, first edition, 1985.

[224] Wie, B., Hu, A., and Singh, R. Multibody interaction effects on space station attitude control and momentum management. *Journal of Guidance, Control and Dynamics*, 13(6):993–999, November-December 1990.

[225] Wilkinson, J. H. *The Algebraic Eigenvalue Problem*. Clarendon Press, 1965.

[226] Willshaw, D. J. and von der Malsburg, C. How patterned neural connections can be set up by self-organization. In *Proceedings of the Royal Society of London B*, volume 194, pages 431–445, 1976.

[227] Wolf, A., Swift, J. B., Swinney, H. L., et al. Determining lyapunov exponents from a time series. *Physica*, 16D:285–317, 1985.

[228] Woo, H. H., Morgan, H. D., and Falangas, E. T. Momentum management and attitude control design for a space station. *Journal of Guidance, Control and Dynamics*, 11(1):19–25, January-February 1988.

[229] Yuan, J. S.-C. and Stieber, M. E. Robust beam-pointing and attitude control of a flexible spacecraft. *Journal of Guidance, Control and Dynamics*, 9(2):228–234, March-April 1986.

Index

action network, 105
activation function, 49
actuation system, 25
Adams-Moulton method, 45
adaptive attitude control, 38
adaptive control architecture, 134
adaptive controller, 13
adaptive critics, 104
adaptive resonance theory (ART), 65
adaptive system, 6
ADP, *see* approximate dynamic programming
angular momentum, 27
angular velocity, 28
ANN, *see* artificial neural networks
approximate dynamic programming (ADP), 104
ART, *see* adaptive resonance theory
ART1, 68
ART2, 68
ART3, 68
artificial neural networks (ANN), 47
ARTMAP, 68
associative memory, 54
attitude control, 25
attitude control approaches, 38
attitude control simulator, 46
attitude determination, 24
attitude maneuver control, 25

attitude prediction, 24
attitude stabilization, 25
automatic program discovery, 119
automatic control, 11
automobile control, 40

backpropagation algorithm, 50
bang-bang control, 41
Bellman equation, 104
Boltzmann machines, 58
bucket brigade algorithm, 119

category resolution, 66
celestial mechanics, 24
certainty equivalence principle, 14
chaos, 8
classifier systems, 118
closure property, 122
comparison layer, 66
competitive learning networks, 64
conflicting objectives, 30
content-addressable memory, 54
continuing methods, 43
control, 11
control inputs, 108
control moment gyros, 40
convergence of numerical methods, 43
critic network, 104
crossover, 112

dART, *see* Distributed ART
depth of individuals, 122
DHP, *see* dual heuristic dynamic programming
difference equations, 6
discrete variable method, 43
discretization error, 43
dispersion, 8
Distributed ART, 68
dual heuristic dynamic programming (DHP), 105
dynamic, 5
dynamic programming, 104
dynamical system, 5

Earth satellite, 25
effector system, 119
effectors, 119
Euler equations, 30
evolution, 112

feature maps, 62
feedback, 11
feedforward networks, 49
first order differential equations, 169
Fourier transform, 7
fractal structure, 73
function set, 121
fundamental matrix, 170
Fusion ARTMAP, 68
Fuzzy ART, 68
Fuzzy ARTMAP, 68
fuzzy logic controller, 40

game theory control, 39
Gaussians, 69
GDHP, *see* global dual heuristic dynamic programming

generalization vs speed dilemma, 167
generalized delta rule, 50
genetic algorithms, 112
    flowchart, 114
    Michigan approach, 112
    Pittsburgh approach, 112
    steady state, 116
genetic programming (GP), 119
global learning networks, 70
globalized dual heuristic dynamic programming (GDHP), 105
GP, *see* genetic programming
GP-B satellite, 40
graph coloring problem, 61
gravity gradient spacecraft, 39
gyro-torquers, 40

HDP, *see* heuristic dynamic programming
heuristic dynamic programming, 105
hidden units, 49
higher order intelligence, 106
Hopfield networks, 52, 54
human pilot model, 41

Ideal performance, 108
ill-posed problems, 99
inertial orientation, 25
inputs, 6
Instructor, 159
Instructor module, 159
intelligent control system, 13
intrinsic parallelism, 117
inverse dynamics, 106
inverse models, 99
Ising model, 55

INDEX

Kalman filter, 40
Kirchoff, 30
Kohonen network evolution, 63
Kohonen networks, 62

laminar motion, 7
Laplace transform, 7
LFRBM equations, 30
limit cycle, 85
linear systems, 7
local discretization error, 44
local learning networks, 70
local linearization, 20
locally predictive nets (LPN), 72
logic system, 25
long term memory (LTM), 68
Lorenz equations, 9
LPN, *see* locally predictive networks
LTM, *see* long term memory
Lyapunov exponents, 78
Lyapunov function, 54
Lyapunov spectrum, 78

Mackey-Glass equation, 73
mammalian brain, 106
Markov chains, 61
McDonnell Douglas Space Systems, 39
Mediator, 108
message list, 118
model reference adaptive control, 16, 102
moment of momentum, 27
momentum saturation, 40
momentum term, 51
Monte Carlo methods, 58
MRAC, *see* model reference adaptive control

MRAC analysis
    gradient approach, 17
    Lyapunov functions, 17
    passivity theory, 17
MRAC parallel, 17
multi-step methods, 43
multilayer perceptrons (MLP), 68
Mutation, 114

natural selection, 111
network paralysis, 76
neural MRAC adaptive control, 102
neuro-genetic control, 134
neurocontrol, 97
neurodynamic programming, 104
Newton, 24
non-spinning spacecrafts, 25
nonhomogeneous system, 169
nonlinear dynamic system, 7
nonlinear observer, 39
nonlinear systems, 7
NP-hard problem, 113
numerical methods, 42

OGY method, 19
one step methods, 43
Online net, 108
Online network, 106
orbit determination, 24
outputs, 6

pattern of change, 5
perceptrons, 49
Pilot, 106
plant, 11
Poincaré section, 87
Prandtl number, 9
prediction, 166

premature convergence, 117, 121
principal axes, 29
principal moments of inertia, 29
protected division, 123
pseudocode
    genetic algorithm selection, 130
    genetic algorithms, 129
    simulated annealing, 60

radial basis function networks, 68
ramped half-and-half method, 123
RBF, 69
receptive fields, 69
recognition layer, 66
redundant systems, 166
reinforcement control, 103
reinforcement learning, 103
repair operators, 113
reproduction, 112
rigid body orientation, 31
round-off error, 43
Runge-Kutta method, 43

Schema theorem, 116
schemata, 116
self-learning controllers, 12
self-tuning regulator (STR), 14
Sensed performance, 108
sensing system, 25
sensitivity analysis, 21
sensitivity on initial conditions, 9
separatrix, 76
series-parallel identification, 72
short term memory (STM), 68
simulated annealing, 58
simultaneous recurrent networks, 70
space navigation, 24
Space Station Freedom, 39

spin glass states, 55
spin stabilization, 25
spin stabilization (stable axis), 139
spin stabilization (unstable axis), 143
SPOT satellite, 42
spurious stable states, 55
SRN, *see* simultaneous recurrent networks
stability-plasticity dilemma, 64
stable state, 106
starting methods, *see* 1-step methods
state variables, 6
steepest descent methods, 50
STM, *see* short term memory
stochastic networks, 58
STR, *see* self tuning regulator
sufficiency property, 122
superposition principle, 7
supervised control, 98
supervised learning, 60
system, 5

terminal set, 121
thermal equilibrium, 60
three axis stabilization, 25
time-evolution, 5
time-lagged recurrent networks, 70
TLRN, *see* time-lagged recurrent networks
Training net, 109
Traveling Salesman problem (TSP), 61, 113
truncation error, 43
TSP, *see* Traveling Salesman problem
turbulent motion, 7

INDEX

underlying design problem, 15
unpredictable dynamic systems, 8
unstable rotation of a rigid body, 143
unsupervised learning, 62
UOSAT satellite, 41

Van der Pol equation, 82
vehicle dynamics system, 25
vigilance parameter, 65
Voyager, 42

weak control, 21